PHYSICAL COMPONENTS OF TENSORS

CRC Series in

APPLIED and COMPUTATIONAL MECHANICS

Series Editor: J.N. Reddy, *Texas A&M University*

PUBLISHED TITLES

ADVANCED MECHANICS OF CONTINUA
Karan S. Surana

ADVANCED THERMODYNAMICS ENGINEERING, Second Edition
Kalyan Annamalai, Ishwar K. Puri, and Miland Jog

APPLIED FUNCTIONAL ANALYSIS
J. Tinsley Oden and Leszek F. Demkowicz

COMBUSTION SCIENCE AND ENGINEERING
Kalyan Annamalai and Ishwar K. Puri

COMPUTATIONAL MODELING OF POLYMER COMPOSITES: A STUDY OF CREEP AND ENVIRONMENTAL EFFECTS
Samit Roy and J.N. Reddy

CONTINUUM MECHANICS FOR ENGINEERS, Third Edition
Thomas Mase, Ronald Smelser, and George E. Mase

DYNAMICS IN ENGINEERING PRACTICE, Tenth Edition
Dara W. Childs

EXACT SOLUTIONS FOR BUCKLING OF STRUCTURAL MEMBERS
C.M. Wang, C.Y. Wang, and J.N. Reddy

THE FINITE ELEMENT METHOD IN HEAT TRANSFER AND FLUID DYNAMICS, Third Edition
J.N. Reddy and D.K. Gartling

MECHANICS OF MATERIALS
Clarence W. de Silva

MECHANICS OF SOLIDS AND STRUCTURES, Second Edition
Roger T. Fenner and J.N. Reddy

MECHANICS OF LAMINATED COMPOSITE PLATES AND SHELLS: THEORY AND ANALYSIS, Second Edition
J.N. Reddy

MICROMECHANICAL ANALYSIS AND MULTI-SCALE MODELING USING THE VORONOI CELL FINITE ELEMENT METHOD
Somnath Ghosh

NUMERICAL AND ANALYTICAL METHODS WITH MATLAB®
William Bober, Chi-Tay Tsai, and Oren Masory

NUMERICAL AND ANALYTICAL METHODS WITH MATLAB®
FOR ELECTRICAL ENGINEERS
William Bober and Andrew Stevens

PHYSICAL COMPONENTS OF TENSORS
Wolf Altman and Antonio Marmo De Oliveira

PRACTICAL ANALYSIS OF COMPOSITE LAMINATES
J.N. Reddy and Antonio Miravete

SOLVING ORDINARY AND PARTIAL BOUNDARY VALUE PROBLEMS IN
SCIENCE AND ENGINEERING
Karel Rektorys

STRESSES IN BEAMS, PLATES, AND SHELLS, Third Edition
Ansel C. Ugural

PHYSICAL COMPONENTS OF TENSORS

Wolf Altman
Antonio Marmo De Oliveira

CRC Press is an imprint of the
Taylor & Francis Group, an **informa** business

CRC Press
Taylor & Francis Group
6000 Broken Sound Parkway NW, Suite 300
Boca Raton, FL 33487-2742

Version Date: 20140908

International Standard Book Number-13: 978-1-4822-6382-4 (Hardback)

Visit the Taylor & Francis Web site at
http://www.taylorandfrancis.com

and the CRC Press Web site at
http://www.crcpress.com

Contents

Foreword by J. N. Reddy

Analytical descriptions of physical phenomena using the laws of nature should be independent of the choice of a coordinate system. Therefore, we may seek to represent the laws in a manner independent of a particular coordinate system. A way of doing this is provided by tensors. The use of tensor notation in formulating natural laws not only leaves them *invariant* to coordinate transformations but also leads to a deeper understanding of the problem.

The book *Physical Components of Tensors with Application to Problems in Applied Mechanics* by Wolf Altman and Antonio Marmo de Oliveira is a welcome addition to the literature on mathematical physics. The book is unique in the sense that it contains an authoritative and comprehensive account of tensor calculus that is based on transformations of bases of vector spaces rather than on transformations of coordinates. The book treats physical and nonholonomic components of tensors with application to the theories of elasticity and shells. The theory as it was developed shows the relationship and compatibility among several existing definitions of physical components of tensors when referred to nonorthogonal coordinates. Thus, *Physical Components of Tensors* is a textbook that clarifies the important aspects of tensor calculus, emphasizing its most important practical aspects. For this reason, it is the best book available on the market. As such it belongs on the bookshelves of every technical library, and in the hands of every graduate student or engineer seriously interested in solid and structural mechanics (e.g., rods, plates, and shells), and should become a widely used textbook or reference book in graduate level courses.

Few words concerning the senior author are in order (summaries of professional highlights of the authors can be found in the section *Biographies of the Authors*). Wolf Altman received his master's degree in 1964 and his Ph.D. in 1966 under the supervision of Prof. D. H. Young at Stanford University. After returning to Brazil, Wolf's research has developed in two fronts: (i) *Weak formulations* including the effects of conservative and nonconservative loads of circulating type that can be applied to problems of beams, plates, and shells.

With different considerations of the order of magnitude of the deformations and rotations, those formulations allow us to obtain several refined theories of shells, including the effect of nonconservative loads. (ii) *Physical and nonholonomic components of tensors* with application to nonlinear theories of shells. The following new aspects are tackled: concepts of nonholonomic components of tensors of double field; the connection between physical and nonholonomic components viewed as an isomorphism between vector spaces. In this front, the theory is applied to problems representing the theories of elasticity and of shells, such as the counterparts of the nonholonomic equations of the theories in question. It is important to mention the pioneering work *Physical Components of Tensors* published in the prestigious journal Tensor NS in 1978. As a consequence of this work, two invited articles were published on the commemorative volume in honor of the 80 years of the great Japanese mathematician Akitsugu Kawaguchi. The research activities of this phase, which certainly will go on for many years, are materialized in 30 articles published from 1978 to this date, most of them in international journals and followed up by many citations from other authors. In many of these works, he had the valuable collaboration from his colleague and former doctoral student Marmo de Oliveira as well as his other students.

In closing it should be mentioned that Wolf Altman is a great human being and an unconditional friend of his friends, students, colleagues, and collaborators. He provides encouragement and mentorship to young researchers with promising academic careers, and he is a constant admirer of the highest evidences of the human spirit. The fact that he authored this book at an age of 88 speaks amply for his passion for technical work and, more importantly, for tensor calculus. It is a privilege and honor to know Wolf Altman and for being considered as his friend.

J. N. Reddy
College Station, Texas

"[He][1] is one of the pioneers of engineering research in Brazil, a natural scientist, someone passionate for the adventure of discovery and invention, with broad view and horizons that transcend his expertise."

Luís Bevilacqua
Rio de Janeiro, Brazil

[1]This statement was published in a volume commemorating Wolf Altman's 70th birthday.

Preface

The analysis of problems involving continuum media is based on models whose formulation and profitable study demand the understanding and skill of tensor calculus. Anyone who lacks the knowledge of this mathematical tool is at a disadvantage in what concerns working effectively in this as well as several other fields of pure and applied mathematics. The theoretical and experimental approach to the motions and other physical processes involve three-dimensional geometry, dynamic loading conditions, and interactions that require a framework provided by adequate coordinate systems. For the most part, all these demands end up in models represented by differential equations whose unknowns are components of tensors.

This book was written for graduate students, professors, and researchers in the areas of elasticity and shell theories. It contains a comprehensive account of tensor calculus (in both holonomic and anholonomic coordinate systems) that is based rather on transformations of bases of vector spaces than on transformations of coordinates, since the latter are not always available. It also establishes a theory of physical and anholonomic components of tensors and applies the theory of dimensional analysis to tensors and (anholonomic) connections. The notion of physical components of tensors enables the homogeneity in the dimensional analysis in any system of coordinates and it is rather the searching for components with this homogeneity property that leads to the concept of physical and anholonomic components for tensors. So, this expanded tensor calculus tries to make use of natural (or holonomic) bases, usually denoted by the bold letters **g** or **G**, as well as the anholonomic bases, represented by the bold letters **e**, **e***, **E**, and **E***. The differences and similarities between the natural tensor components of the classical calculus and the physical components of tensors can be listed as follows:

a) The physical components are those resulting from laboratory measurements.

b) The tensor components coincide with the physical components when one uses an orthonormal Cartesian coordinate system.

c) Expressions written in terms of physical components must be homogeneous in the sense that all the terms in each expression have the same physical dimensions. The tensor components do not have this homogeneity property.

d) Both the physical and the tensor components must have an analogous behavior after a change of bases or of coordinates. We can explain this analogy by means of an example (see Chapter 4):

We say that n quantities $T^s(x)$ associated with a point P are the components of a contravariant vector $\mathbf{T}$ if they transform on change of coordinates from x to $\bar{x}$ according to the expression

$$\bar{T}^k(\bar{x}) = \frac{\partial \bar{x}^k}{\partial x^s} T^s(x).$$

Analogously, we say that n quantities $[T^s(x)]_e$ associated with a point P are the physical components of a contravariant vector $\mathbf{T}$ if they transform on change of bases from $\mathbf{e}$ to $\bar{\mathbf{e}}$ (and on change of coordinates from x to $\bar{x}$) according to the expression

$$[\bar{T}^k(\bar{x})]_{\bar{e}} = \left[\frac{\partial \bar{x}^k}{\partial x^s}\right]_{\bar{e}\otimes e} [T^s(x)]_e \,,$$

where the partial derivatives are evaluated at the point P. In Chapter 4 the formula above is generalized for a tensor of order n.

A particular although satisfactory theory of physical components in orthogonal curvilinear coordinates was presented in [McConnell, 1931] and, concerning the general coordinate systems (including nonorthogonal ones), several authors have defined physical components for second order tensors (e.g., [Green and Zerna, 1954], [Synge and Schild, 1978], [Sedov, 1965], and [Truesdell, 1953]). All those definitions are different from each other and agree only in the case of orthogonal coordinates, in which case they are all equivalent to that of McConnell. Usually, to avoid any apparent ambiguity in those definitions, the tensor components are transformed in such a way as to refer to an orthogonal coordinate system and thus allowing the use of McConnell's concept of physical components. A more general theory developed by us (see [Altman and Oliveira, 1977], [Oliveira and Altman, 1981], and [Altman and Oliveira, 1995]), which includes all physical components (mixed, covariant, and contravariant) and is based on the method of invariance of tensor representation, shows the compatibility and relationship among those definitions. In the context where each tensor is seen as a linear transformation between two properly chosen vector spaces, the concept of physical components depends fundamentally on the linear transformation as well as on the bases of the vector spaces in question. Since in the anholonomic case our method is built up considering the bases **e**, **f**, **e***, and **f*** for the deformed configuration (or **E**, **F**, **E***, and **F*** for the undeformed configuration), and thus doubling their number in relation to the holonomic case, it is expected that the results should be different. On the other hand, the method of anholonomic components developed by [Truesdell, 1954] and [Ericksen, 1960] does not include all cases, as acknowledged by Ericksen himself:

"*Possibly a method succinctly general to cover every proposal which might appeal on some or other intuitive grounds could be constructed. In this connection, we mention only that the method of anholonomic components, which appears quite general, does not include Green and Zerna's definition.*"
[Ericksen, 1960]

Truesdell and Ericksen dealt only with anholonomic components of single field referred to the bases **e** and **e*** (or **E** and **E***), therefore omitting the other pair of bases (**f**, **f*** or **F**, **F***). The main reason for introducing the physical components is the need for the "dimensional homogeneity" of the tensor components, that is, every tensor components sharing the same physical dimension. Some authors consider only the physical components that stem from unitary bases. In our case, defining a physical component as the one that has the same physical dimension as that of the tensor, we shall obtain a class of physical components that is more general. Despite not being unitary, the bases **e*** and **f** yield such components.

In this book, we show by means of connections between physical components and anholonomic components that the latter components, defined by

$$[T^{A\,c}]_{F\otimes e} = T^{I\,k} F_I^A e_k^c,$$

are referred to the bases **F** and **e**. Since Truesdell and Ericksen omitted the basis **F**, they could not include the anholonomic counterparts of Green and Zerna's definition in their method. The same problem will be seen in Table 4.2 of Chapter 4, where twelve out of the sixteen components are excluded for not considering the basis **F**. Therefore, considering that we deal with all the above mentioned components in this book, we can say that the theory developed here is more general.

In composing the book, we have assumed a basic knowledge of Linear Algebra and elementary calculus, although all the relevant concepts and results in these subjects are revisited in the first three chapters. Since it was not our intention to present an exhaustive account of the subject, we have touched upon only those topics related to our own contribution to the theory. So, the first three chapters introduce the mathematical backgrounds for the theory that is developed and applied in Chapters 4 and 5. Chapter 1 deals with some basic concepts of Linear Algebra, introducing the vector spaces and the further structures imposed on them by the notions of inner products, norms, and metrics. The fundamental representation theorem for finite-dimensional vector spaces and the concepts of Cartesian and curvilinear systems of coordinates as well as transformation of coordinate systems are presented. Chapter 2 focusses on the main algebraic operations for vectors and tensors and also on the notions of duality, tensor products, and component representation of tensors. Chapter 3 presents the classical tensor calculus that functions as the advanced prerequisite for the developments of the subsequent chapters. In particular, covariant differentiation of vectors, Christoffel symbols, and their metric tensors are introduced, and the fundamental integral theorems of Gauss and of Stokes are reproduced for reference.

Chapter 4 presents the theory of physical and anholonomic components of tensors by associating them to the spaces of linear transformations and of tensor products. In this context, formulas are obtained to describe a more general tensor calculus than the one that is developed in the existing books in the sense that it includes as a particular case the tensor calculus based on coordinate transformations. Also, explicit expressions for the usual connections are obtained as well as expressions for the anholonomic components of the covariant derivative of a tensor. Since the classical theory is built up by considering only the holonomic bases for the required calculations, there is no need to specify these bases when denoting the holonomic and physical components of tensors. However, since the anholonic tensor calculus depends on which holonomic and anholonomic bases are considered, we found it necessary to make the specifications of each basis explicit in the notation for the components of tensors. This chapter also shows that the concept of physical components of tensors depends fundamentally on the bases of a space of linear transformations and this is done by showing how the physical components vary with coordinate transformations.

Chapter 5 advances two applications of this theory. The first one is concerned with a continuous media by focussing on deriving a system of partial differential equations for the stress tensor for the equilibrium under the action of forces. The second application is on the theory of deformation of elastic shells. We start with a characterization of shells and give a summary of governing equations, both in terms of tensor components and of physical components.

This book has been a long time in the writing and a number of persons have been extremely valuable throughout the process. We are grateful to

Luís Bevilacqua, who pointed out the opportunity to present our ideas in a book, and to Paulo Ivo Braga de Queiroz for reading the drafts and giving suggestions. We are grateful also to three late friends, Guilherme de La Penha and Fernando Venâncio Filho, for many profitable discussions, and Carlos Alberto Borges, for his encouragements at the early stages of the process. We would like to thank Leo Huet Amaral for his valuable suggestions on the chapters concerning the mathematical preliminaries and for the useful insights that we have gathered from reading his book on linear algebra [Amaral, 2002]. Marcos A. Botelho helped to bring the book into form and we are indebted for his participation. Our gratitude goes also to Ariovaldo Felix Palmerio for his support in the final steps of the process. Finally, the authors gratefully acknowledge the encouragement and many hours of effort put in fine-tuning the manuscript by J. N. Reddy.

We are in debt to our families and our wives, Sara and Mércia, whose support and understanding were important during the demands of energy that we have faced while working hard on this book.

The typesetting of this book was made using the Preparation Document System LaTeX by Marcos A. Botelho, Leo H. Amaral, and P. I. Braga de Queiroz.

Wolf Altman
Antonio Marmo de Oliveira
São Paulo and Taubaté, Brazil

Biographies of the Authors

WOLF ALTMAN obtained his Ph.D. from Stanford University in 1966 under the supervision of Professor D. H. Young. His dissertation was titled *Stresses in Conical Shells*, which was published in the *Journal of IABSE* and became a classic. Professor Altman is an engineering educator and researcher with experience as a consultant in structural mechanics in general and in the City Hall of São Caetano do Sul. His research endeavors, which include several graduate student supervisions and over 60 articles in international journals, have been mainly on weak variational formulations, including the effects of conservative and nonconservative loads on beams, plates and shells, and on physical and nonholonomic components of tensors with application to the theory of elasticity and shells. He was one of the organizers of the first Pan American Congress of Applied Mechanics, having become a constant paper contributor to its subsequent conferences. In 1996, the Brazilian academic community organized a scientific congress in tribute of his 70th birthday.

ANTONIO MARMO DE OLIVEIRA earned his doctorate at the Instituto Tecnológico de Aeronáutica in 1977 under the supervision of Professor Wolf Altman. He was full professor at Instituto Tecnológico de Aeronáutica and University of Taubaté until his retirement, having lectured and supervised master and doctorate students in a number of universities in Brazil during this period. He performed research in the broad fields of mechanics, applied mathematics (variational and tensor calculus), and engineering science (composite structures), and he has published 9 books and over 50 articles in journals and magazines as well as 300 chronicles in Taubaté's newspapers. He was awarded the 1966 Esso Prize of Sciences and some commendations in Taubaté city, where he was the head of the university during 2000–2002. He is member of the American Mathematical Society, the Tensor Society (Chigasaki, Japan), the Brazilian Society of Applied and Computational Mathematics (SBMAC), and the Brazilian Institute of Composites (IBCOM). Currently he works as a consultant for reinforced industrial plastics.

Biographies of the Authors

List of Symbols

Here we present a list of the main symbols. First, on this page, we present symbols used in the equations of shells. Then, we present a list of other symbols used in the book. The list is not extensive for all the symbols used throughout the book.

$\mathbf{M}^N = \mathbf{M}^N(\theta^\alpha, t; \mathbf{v})$	Stress-resultant couple vectors of order $N \geq 0$
$\mathbf{M}^{n\alpha}$	Stress couples of order $n \geq 0$
$M^{\alpha i}$	Components of the contact director couple $\mathbf{M}$
$\mathbf{N} = \mathbf{N}(\theta^\alpha, t; \mathbf{v})$	Stress-resultant force
$\bar{\boldsymbol{l}}^n$	Body force resultants of order $n \geq 0$
$\mathbf{m}$	Surface director couple measured per unit area in the deformed configuration
$\mathbf{m}^n$	Shear stress resultants of order $n \geq 0$
$\mathbf{a}_3$	Unit normal vector in the deformed configuration
$\mathbf{A}_3$	Unit normal vector in the reference configuration
$\mathbf{a}_\alpha,\ \mathbf{A}_\alpha$	Base vectors of a surface in the deformed and undeformed configurations
B	Flexural rigidity
$b_{\alpha\beta},\ B_{\alpha\beta}$	Second fundamental form of a surface in the deformed and undeformed configurations

$\mathbf{g} = \{\mathbf{g}_1, \mathbf{g}_2, \ldots, \mathbf{g}_n\}$	Holonomic basis
$\mathbf{g}^* = \{\mathbf{g}^1, \ldots, \mathbf{g}^n\}$	Reciprocal basis
$(v_1, \ldots, v_n)$	Covariant components
$(v^1, \ldots, v^n)$	Contravariant components
$g_{ij} = \mathbf{g}_i \cdot \mathbf{g}_j$	Metric tensor
$\delta^i_j = \begin{cases} 1, & \text{if } i = j \\ 0, & \text{if } i \neq j \end{cases}$	Mixed components of the unit second-order tensor
$J^j_{\bar{j}} = \dfrac{\partial x^i}{\partial \bar{x}^j}$	Entries of the Jacobian matrix J
$N^{\bar{j}}_i = \dfrac{\partial \bar{x}^j}{\partial x^i}$	Entries of the Jacobian inverse $N = J^{-1}$
$T^{\bar{i}}_{\bar{j}} = T^i_j N^{\bar{i}}_i J^j_{\bar{j}}$	Transformation for the mixed components
$T^{\bar{i}\bar{j}} = T^{ij} N^{\bar{i}}_i N^{\bar{j}}_j$	Transformation for the contravariant components
$L(V, W)$	Space of all linear transformations from V into W
$V^* = L(V, \mathbb{R})$	Dual space of V
$\mathbf{u}^*, \mathbf{v}^*, \mathbf{w}^*$	Elements of V^*
$T^i_k = [T^i_k]^f_e$	Components of the linear transformation T, with $T(\mathbf{e}_k) = T^i_k \mathbf{f}_i$
$V \otimes U$	Tensor product of the spaces V and U
$\mathbf{e}_i \otimes \mathbf{f}_j$	Elements of a basis for $V \otimes U$
$\mathbf{T} = T^{ij} \mathbf{g}_i \otimes \mathbf{g}_j$	Decomposition of a second-order tensor T in a base product $\mathbf{g}_i \otimes \mathbf{g}_j$
$\mathbf{T} = T^{i_1 \cdots i_p} \mathbf{e}_{i_1} \otimes \cdots \otimes \mathbf{e}_{i_p}$	Contravariant tensor $\mathbf{T}$ of order p
$\mathbf{S} = S_{j_1 \cdots j_q} \mathbf{e}^{j_1} \otimes \cdots \otimes \mathbf{e}^{j_q}$	Covariant tensor $\mathbf{S}$ of order q
$\mathbf{A} = \mathbf{T} \otimes \mathbf{S}$	Mixed tensor $\mathbf{A}$ of order $p + q$

Chapter 1

Finite-Dimensional Vector Spaces

The notion of vector spaces essentially provides a framework in which the principle of superposition can be formalized. Once we have this structure, we can move on to consider additional operations to supply it with both geometric and analytic structures. The so-called Euclidean vector space to be defined in the sequel will be one of such geometrically structured spaces.

1.1 Vector spaces and subspaces

The algebraic structure known as vector space is fundamental in many applications. Essentially, the abstraction that is the concept of vector space started off from a formalization of the principle of superposition in Physics. Its structure is built by defining two operations, called the addition of vectors and the multiplication of a vector by a scalar, and depends upon a preliminary choice of a more elementary algebraic structure, called a *field.*

Definition 1.1 By a *field* $\mathbb{F}$ we mean a set $\mathbb{F}$ with at least two elements and two operations defined on $\mathbb{F}$ (one, called *addition,* assigns to every ordered pair (x, y) of elements of $\mathbb{F}$ another element of $\mathbb{F}$, denoted $x + y$; the other, called *multiplication*, also assigns to every ordered pair (x, y) of elements of $\mathbb{F}$ another element of $\mathbb{F}$, denoted xy) satisfying the following properties:

(R1) $(x + y) + z = x + (y + z)$, for all $x, y, z \in \mathbb{F}$.
(R2) $x + y = y + x$, for all $x, y \in \mathbb{F}$.
(R3) There is an element $0 \in \mathbb{F}$, called the *additive identity*, such that

$$x + 0 = 0 + x = x, \forall x \in \mathbb{F}.$$

(R4) To each $x \in \mathbb{F}$ there corresponds an element $-x \in \mathbb{F}$ such that

$$x + (-x) = (-x) + x = 0.$$

(R5) $(xy)z = x(yz)$, for all $x, y, z \in \mathbb{F}$.

(R6) $xy = yx$, for all $x, y \in \mathbb{F}$ (we say that $\mathbb{F}$ is commutative).

(R7) There is an element $1 \in \mathbb{F}$, called the *multiplicative identity*, such that

$$x1 = 1x = x, \forall x \in \mathbb{F}.$$

(R8) For each element $x \in \mathbb{F}$, such that $x \neq 0$, there exists an element $x^{-1} \in \mathbb{F}$, which we call *inverse of x*, such that

$$xx^{-1} = x^{-1}x = 1.$$

(R9) $(x + y)z = xz + yz, \quad$ for all $x, y, z \in \mathbb{F}$.

(R10) $x(y + z) = xy + xz, \quad$ for all $x, y, z \in \mathbb{F}$.

It follows that the identities of each operation, namely the ones denoted as 0 and 1, are unique and that 0, the identity for addition, is always a "zero for multiplication" in the sense that $0x = x0 = 0$ for all $x \in \mathbb{F}$. Three very important examples of fields are the set of rational numbers $\mathbb{Q}$, the set of real numbers $\mathbb{R}$, and the set of complex numbers $\mathbb{C}$, together with their respective usual operations of addition and multiplication that qualify each one of them as a field. For the purposes of this book, only the field of the real numbers will be of interest. We assume that the reader is acquainted with the concept of real numbers, as well as of the other numbers (the natural numbers — elements in $\mathbb{N}$ —, the integers — elements in $\mathbb{Z}$ —, and the rational, irrational, and complex numbers). For those interested in an axiomatic and more comprehensive treatment of real numbers and their subclasses, we refer to [Amaral, 2002].

Going one step further in building algebraic structures, we have the notion of a vector space, as defined next.

Definition 1.2 Let V be a nonempty set and $\mathbb{F}$ be a field. We say that V is a **vector space** (over the field $\mathbb{F}$) if the following axioms are satisfied:

(a) There exists a binary operation, called *addition* and denoted by $+$, assigning to every ordered pair (x, y) of elements of V another element of V, denoted $x + y$, such that

a1. $(a + b) + c = a + (b + c)$, for all a, b, c in V.

a2. There exists an element $0 \in V$ with the property that $a + 0 = 0 + a = a$, for all $a \in V$.

a3. For every $a \in V$ there exists $-a \in V$ such that $a + (-a) = (-a) + a = 0$.

a4. $a + b = b + a$, for all a, b in V.

(b) There exists an operation, called *multiplication by a scalar*, in which every λ in $\mathbb{F}$ can be combined with every a in V to give an element λa in V such that

b1. $\lambda(\mu a) = (\lambda\mu)a$, for all $\lambda, \mu \in \mathbb{F}$ and all $a \in V$.
b2. $(\lambda + \mu)a = \lambda a + \mu a$, for all $\lambda, \mu \in \mathbb{F}$ and all $a \in V$.
b3. $\lambda(a + b) = \lambda a + \lambda b$, for all $\lambda, \mu \in \mathbb{F}$ and all $a \in V$.
b4. $1.a = a$, for all $a \in V$. (Here, 1 denotes the identity element of the field $\mathbb{F}$.)

Axioms a1–a3 say that V supplied with the operation $+$ is an algebraic structure called *group*, and the extra axiom a4 says that the group is commutative or that it is an *Abelian group*. Every element of V is called a *vector*, and we call *scalar* every element of the field. It is usual to denote a vector space simply as V, leaving out the references to the operations and to the field. When the field is that of the real numbers, we say that V is a *real vector space*. Scalars are usually not elements of V but one can easily verify that every field $\mathbb{F}$ is a vector space over itself with the defined operations.

Here are some relevant examples of vector spaces:

1. The vector spaces $\mathbb{F}^n$ and $\mathbb{R}^n$: For any field $\mathbb{F}$ and $n \in \mathbb{N}$, let

$$V = \underbrace{\mathbb{F} \times \mathbb{F} \times \cdots \times \mathbb{F}}_{n \text{ times}} = \mathbb{F}^n$$

be the set of all ordered n-tuples $\mathbf{x} = (x_1, x_2, \ldots, x_n)$ of elements of $\mathbb{F}$ with addition of two n-tuples defined by

$$\begin{aligned} \mathbf{x} + \mathbf{y} &= (x_1, x_2, \cdots, x_n) + (y_1, y_2, \cdots, y_n) \\ &= (x_1 + y_1, x_2 + y_2, \cdots, x_n + y_n) \end{aligned}$$

and multiplication by a scalar given by

$$\lambda \mathbf{x} = (\lambda x_1, \lambda x_2, \cdots, \lambda x_n).$$

Then, the set $V = \mathbb{F}^n$, supplied with the structure provided by these two operations, can be proved to satisfy the axioms a1–a4 and b1–b4 and therefore is an example of a vector space over the field $\mathbb{F}$. Actually, this example is remarkably important because an extensive class of vector spaces, that of all n-dimensional vector spaces over the field $\mathbb{F}$, can be identified with $\mathbb{F}^n$ in a sense that we shall make precise later on. Since throughout this book we shall be dealing with the field of real numbers, from now on we assume $\mathbb{F} = \mathbb{R}$, unless otherwise stated. Thus, in particular, the present example reduces to the vector space $V = \mathbb{R}^n$ over the field $\mathbb{R}$.

2. Given any vector space V over a field $\mathbb{F}$, a mapping $T : V \to V$ is called a *linear transformation* if for any $x, y \in V$ and $\lambda \in \mathbb{F}$,

$$T(x + y) = T(x) + T(y) \text{ and } T(\lambda x) = \lambda T(x).$$

One can show that the set $L(V)$ of all linear transformations from V into V with the operations defined by

$$(T + M)(x) = T(x) + M(x), \forall T, M \in L(V) \text{ and } x \in V$$
$$(\lambda T)x = T(\lambda x), \forall T \in L(V) \text{ and } \lambda \in \mathbb{F}.$$

is itself a vector space (over $\mathbb{F}$).

3. The set of all $n \times n$ real matrices, with $+$ meaning to add the corresponding entries and the multiplication by a real number λ meaning the multiplication of each entry by λ, is a real vector space that we shall denote $\mathbb{R}^{n \times n}$.

4. Let P_n denote the set of real polynomials of degree less than or equal to n, that is, P_n is the set of all polynomials of the form $a_o + a_1 x + \ldots + a_n x^n$, where $a_o, a_1, \cdots, a_n \in \mathbb{R}$. Then, P_n forms a vector space over $\mathbb{R}$ if addition and a scalar multiplication by a real number are defined by

$$(a_o + \cdots + a_n x^n) + (b_o + \cdots + b_n x^n) = (a_o + b_o) + \cdots + (a_n + b_n) x^n$$
$$\lambda(a_o + \cdots + a_n x^n) = \lambda a_o + \cdots + \lambda a_n x^x$$

for all $\lambda, a_i, b_i \in \mathbb{R}$.

5. Consider the set of complex numbers $\mathbb{C}$ with the usual definitions of addition and multiplication by a real number, namely

$$(a + b\,i) + (c + d\,i) = (a + c) + (b + d)i,$$
$$\lambda(a + b\,i) = \lambda a + \lambda b\,i \text{ for all } \lambda, a, b, c, d \in \mathbb{R}.$$

One can verify that $\mathbb{C}$ forms a vector space over the field $\mathbb{R}$. On the other hand, as a consequence of our first example, we also have that $\mathbb{C}$, imbued with the same operations but now taking $\mathbb{C}$ as the scalar field (that is, taken $a, b, c, d \in \mathbb{R}$ but $\lambda \in \mathbb{C}$ above), is another vector space — namely, the vector space $\mathbb{C}$ over the field $\mathbb{C}$. Just take $\mathbb{F}^n$ with $\mathbb{F} = \mathbb{C}$ and $n = 1$ in the first example. But it is interesting to note that, knowing that $\mathbb{C}$ is ultimately $\mathbb{R}^2$ imbued with a structure of field and taking $V = \mathbb{C} = \mathbb{R}^2$ and $\mathbb{F} = \mathbb{R}$ in the first example, to the set of complex numbers can be given two different structures of vector spaces: one being $\mathbb{C}$ as a vector space over the field $\mathbb{C}$ and the other being $\mathbb{C}$ as a vector space over the field $\mathbb{R}$.

Given a vector space V, a very important example of a new vector space induced by V is that of a subset $M \subset V$ which is itself a vector space with the operations inherited from V. More precisely, we have the following

Definition 1.3 Let V be a vector space (over a field $\mathbb{F}$). A *nonempty* subset M of V is a **subspace** of V iff (if and only if) the two conditions are satisfied:

(a) If $\mathbf{u}, \mathbf{v} \in M$ then $\mathbf{u} + \mathbf{v} \in M$.
(b) If $\mathbf{u} \in M$ then $\lambda \mathbf{u} \in M$ for all $\lambda \in \mathbb{F}$.

Some authors prefer to reinforce the condition of nonemptiness by replacing it for the condition $0 \in M$. It is obvious that the nonemptiness of $M \subset V$ together with the axioms (a) and (b) imply that $0 \in M$.

Notice that this definition is equivalent to the fact that a subspace M is a subset which is itself a vector space with the operations inherited from V, since one can easily see that the vector space axioms hold for M because they hold for V.

The following are examples of subspaces, which one can check out straightway. From now on, we shall sometimes omit the phrase "over the field $\mathbb{F}$" because all the vector spaces that we shall consider in this book will be real vector spaces. Thus we might simply say "vector space" instead of "vector space over the field $\mathbb{R}$" or "real vector space."

(1) The subset of the vector space $\mathbb{R}^n$ of all n-tuples of the form

$$(0, x_2, x_3, \cdots, x_n)$$

is a subspace of $\mathbb{R}^n$.
(2) The set consisting only of the zero vector 0 is a subspace of V.
(3) Any vector space V is a subspace of itself.

Example 1.1 Let V be a vector space and M a nonempty subset of V. Prove that the following statements are equivalent:

(a) M is a subspace of V.
(b) For all $\alpha \in \mathbb{R}$ and $u, v \in M$ we have that $u + v \in M$ and $\alpha u \in M$.
(c) For all $\alpha, \beta \in \mathbb{R}$ and $u, v \in M$ we have that $\alpha u + \beta v \in M$.

Solution: That (a) $\Rightarrow$ (b) is immediate from the definition. To show that (b) $\Rightarrow$ (c) note that from (b) it follows that $\alpha\mathbf{u}, \beta\mathbf{v} \in M$ and $\alpha\mathbf{u} + \beta\mathbf{v} \in M$.

Now, for any $\mathbf{u}, \mathbf{v} \in M$, (c) gives that $\mathbf{u} + \mathbf{v} = 1 \cdot \mathbf{u} + 1 \cdot \mathbf{v} \in M$ and that, for any $\lambda \in \mathbb{R}$, $\lambda \mathbf{u} = \lambda \mathbf{u} + 0\mathbf{v} \in M$.

Let V be a vector space and let $S = \{\mathbf{v}_1, \mathbf{v}_2, \ldots, \mathbf{v}_n\} \subset V$ be any finite subset of V. We introduce the notion of a subspace generated (or spanned) by S. We recall that in V any superposition of scaled vectors

$$\lambda_1 \mathbf{v}_1 + \lambda_2 \mathbf{v}_2 + \cdots + \lambda_n \mathbf{v}_n, \tag{1.1}$$

with λ_i being scalars, is called a **linear combination** in V and the scalars $\lambda_1, \lambda_2, \ldots, \lambda_n$, on that order, are called the **components** of the linear combination.

The *set of all linear combinations* (1.1) of the elements $\mathbf{v}_1, \mathbf{v}_2, \ldots, \mathbf{v}_n \in V$ is denoted

$$span\{\mathbf{v}_1, \mathbf{v}_2, \ldots, \mathbf{v}_n\} \text{ or } span\ S \text{ or } [S].$$

The fundamental property of $span\{\mathbf{v}_1, \mathbf{v}_2, \ldots, \mathbf{v}_n\}$ is, as one can easily check, that:

(a) it is a subspace of V and

(b) in fact, it is the smallest subspace of V which contains the elements $\mathbf{v}_1, \mathbf{v}_2, \ldots, \mathbf{v}_n$, in the sense that if M is any subspace of V which contains $\mathbf{v}_1, \mathbf{v}_2, \ldots, \mathbf{v}_n$, then $M \supseteq span\{\mathbf{v}_1, \mathbf{v}_2, \ldots, \mathbf{v}_n\}$.

Definition 1.4 Given a vector space V, consider $S = \{\mathbf{v}_1, \mathbf{v}_2, \ldots, \mathbf{v}_n\} \subset V$. We say that $span\{\mathbf{v}_1, \mathbf{v}_2, \ldots, \mathbf{v}_n\}$ is the **subspace of** V **generated** (or spanned) by the vectors $\mathbf{v}_1, \mathbf{v}_2, \ldots, \mathbf{v}_n$ or by the set S and that S is the **generator** *of* $[S]$. Alternatively, we say that $span\{\mathbf{v}_1, \mathbf{v}_2, \ldots, \mathbf{v}_n\}$ is the **subspace of** V **spanned** by the vectors $\mathbf{v}_1, \mathbf{v}_2, \ldots, \mathbf{v}_n$.

Example 1.2 (a) Prove that $span\{v_1, v_2, \ldots, v_n\}$ is a subspace of V. (b) (Space spanned by different sets) Consider the space $\mathbb{R}^3$ and the following subsets:

$$S = \{(1,2,1), (3,1,5)\}, \quad W = \{(1,2,1), (3,1,5), (3,-4,7)\}.$$

Are $[S]$ and $[W]$ identical subspaces of $\mathbb{R}^3$?

Solution: (a) Given any linear combinations $\mathbf{u} = \sum_{i=1}^{n} c_i \mathbf{v}_i$ and $\mathbf{v} = \sum_{i=1}^{n} d_i \mathbf{v}_i$ of these n vectors, we have that

$$\mathbf{u} + \mathbf{v} = \sum_{i=1}^{n} c_i \mathbf{v}_i + \sum_{i=1}^{n} d_i \mathbf{v}_i = \sum_{i=1}^{n} (c_i + d_i)\mathbf{v}_i.$$

On the other hand, for any $\lambda \in \mathbb{R}$,

$$\lambda \mathbf{u} = \sum_{i=1}^{n} c_i \mathbf{v}_i = \sum_{i=1}^{m} \lambda c_i \mathbf{v}_i.$$

Therefore, $\mathbf{u}+\mathbf{v}$ and $\lambda\mathbf{u}$ are linear combinations of $\{\mathbf{v}_1, \mathbf{v}_2, \ldots, \mathbf{v}_n\}$. Therefore, they belong to $span\{\mathbf{v}_1, \mathbf{v}_2, \ldots, \mathbf{v}_n\}$.

(b) Since $(3, -4, 7) = -3.(1, 2, 1) + 2.(3, 1, 5)$ we may write:

$$\begin{aligned}
[W] &= \{a(1, 2, 1) + b(3, 1, 5) + c(3, -4, 7);\ a, b, c \in \mathbb{R}\} \\
&= \{(a - 3c)(1, 2, 1) + (b + 2c)(3, 1, 5);\ a, b, c \in \mathbb{R}\} \\
&= \{m(1, 2, 1) + n(3, 1, 5);\ m, n \in \mathbb{R}\} \\
&= [S].
\end{aligned}$$

There is a cast of concepts that depend on the notion of linear combination and are crucial for the study of vector spaces. They are the concepts of linear dependence and linear independence for a set of vectors, and the concepts of basis and dimension of a vector space. We shall focus on them next.

1.2 Basis of a vector space

Definition 1.5 Let V be a vector space over $\mathbb{R}$. A set of vectors

$$\{\mathbf{g}_1, \mathbf{g}_2, \ldots, \mathbf{g}_n\} \subset V$$

is said to be **linearly dependent** iff there exist scalars $\lambda_i \in \mathbb{R}, i = 1, \cdots, n$, not all zero, such that

$$\lambda_1 \mathbf{g}_1 + \lambda_2 \mathbf{g}_2 + \cdots + \lambda_n \mathbf{g}_n = 0. \tag{1.2}$$

Otherwise, if $\lambda_1 \mathbf{g}_1 + \lambda_2 \mathbf{g}_2 + \cdots + \lambda_n \mathbf{g}_n = 0$ implies $\lambda_1 = \lambda_2 = \cdots = \lambda_n = 0$, we say that the vectors $\mathbf{g}_1, \mathbf{g}_2, \cdots, \mathbf{g}_n$ are **linearly independent**.

Definition 1.6 We say that a vector space V is n-**dimensional**, or that $\dim V = n$, iff there exists a set $\{\mathbf{v}_1, \mathbf{v}_2, \ldots, \mathbf{v}_n\}$ of n linearly independent vectors such that

$$span\{\mathbf{v}_1, \mathbf{v}_2, \ldots, \mathbf{v}_n\} = V.$$

Any set with this property is called a **basis** of V. If this is not the case, that is, if the vector space V can not be spanned by a finite number of linearly independent vectors, we say that the dimension of V is infinite, or that V is an infinite-dimensional vector space.

It is worth noting that one can prove the important fact that any basis of a vector space of dimension n *has exactly* n elements. As an example, we show that the set $\{\mathbf{m}_1, \ldots, \mathbf{m}_n\}$, with

$$\begin{aligned} \mathbf{m}_1 &= (1, 0, 0, \ldots, 0), \\ \mathbf{m}_2 &= (0, 1, 0, \ldots, 0), \\ \mathbf{m}_n &= (0, 0, \ldots, 0, 1), \end{aligned}$$

is a basis of the vector space $\mathbb{R}^n$ over the field $\mathbb{R}$. Indeed, given any $(x_1, \ldots, x_n) \in \mathbb{R}^n$, we can write

$$\mathbf{x} = x_1\mathbf{m}_1 + \cdots x_n\mathbf{m}_n$$

so that $span\{\mathbf{m}_1, \ldots, \mathbf{m}_n\} = \mathbb{R}^n$. Also,

$$\lambda_1\mathbf{m}_1 + \cdots + \lambda_n\mathbf{m}_n = 0$$

implies $(\lambda_1, \lambda_2, \ldots, \lambda_n) = (0, 0, \ldots, 0)$, showing that $\{\mathbf{m}_1, \ldots, \mathbf{m}_n\}$ is linearly independent.

Obviously, given a basis with n elements, any set with $n + 1$ vectors is linearly dependent. A vector space which has a finite number of linearly independent vectors that generate it (in other words, a basis in the sense above) is called a finite-dimensional vector space. If there is no bound for the number of linearly independent vectors, then the vector space is said to be infinite dimensional. In what follows, we shall be dealing with finite-dimensional vector spaces.

Given a basis $\mathbf{g} = \{\mathbf{g}_1, \mathbf{g}_2, \ldots, \mathbf{g}_n\}$ in a n-dimensional vector space V, the fact that the basis spans the vector space implies that any vector $\mathbf{v}$ of V can be expressed as a linear combination of the elements of that basis. By this, we mean that there exist[1] scalars $v^1, v^2, \ldots, v^n$ such that

$$\mathbf{v} = v^1\mathbf{g}_1 + v^2\mathbf{g}_2 \cdots + v^n\mathbf{g}_n = \sum_{k=1}^{n} v^k\mathbf{g}_k.$$

[1]We note that the superscripts do not stand for powers but are used to distinguish the components of v.

We refer to the n-tuple $(v^1, \ldots, v^n)$ as the **components of v with respect to the basis g**.

As a means of labeling the elements in a collection of quantities we can use the *index notation* (recall that an index is a letter which appears as a subscript or superscript) [Reddy and Rasmussen, 1982; Reddy, 2013]. We can illustrate this notation as follows. For $k \in \{1, 2, ..., n\}$, we denote by A^k the kth component of vector $\mathbf{A}$ and by $\mathbf{g}_k$ the kth vector of the basis $\mathbf{g} = (\mathbf{g}_1, \mathbf{g}_2, \ldots, \mathbf{g}_n)$, now presented as an ordered n-tuple to make precise the order of the vectors for reference. Thus, if $\mathbf{x}$ and $\mathbf{y}$ have components $(x^1, \ldots, x^n)$ and $(y^1, \ldots, y^n)$ with respect to the same basis $(\mathbf{g}_1, \mathbf{g}_2, \ldots, \mathbf{g}_n)$, then the components of $\mathbf{x} + \mathbf{y}$ relative to this basis are $(x^1 + y^1, \ldots, x^n + y^n)$. Similarly, $(\lambda x^1, \ldots, \lambda x^n)$ are the components of λx, for any scalar λ.

Finally, in order to simplify the use of index notation in the summations that appear when considering linear combinations like

$$v^1\mathbf{g}_1 + v^2\mathbf{g}_2 \cdots + v^n\mathbf{g}_n = \sum_{k=1}^{n} v^k\mathbf{g}_k,$$

we adopt a *summation convention* by assuming that any term in which an index is repeated denotes a sum of all terms obtained by assigning numbers $1, 2, \ldots, n$ in turn to the repeated indexes. According to this convention, the linear combination above can be denoted as

$$v^1\mathbf{g}_1 + v^2\mathbf{g}_2 \cdots + v^n\mathbf{g}_n = v^k\mathbf{g}_k,$$

so dispensing with the sum signal $\sum_{k=1}^{n}$. However, this convention will not apply whenever we have double indices. So, whilst $v^k\mathbf{g}_k$ means the sum of vectors above according to the summation convention, an expression like $v^{kk}\mathbf{g}_k$ will not mean a sum but just one single, say $\mathbf{v}_k$, vector parameterized by k.

1.3 Inner products, norms, and metrics

It is common to designate by the name *space* any set endowed with a particular structure (that is, by a structure we mean a function), usually referred to as an operation, relating the elements of the set. For instance, we name *inner product space* any vector space endowed with an additional specific structure (an operation called inner product) that gives it a geometric feature by allowing us to define the concepts of angle between their elements and length of an element. These concepts are derived in a natural way from the inner product. A *normed space* is a vector space endowed with a structure (a function called *norm*) that allows us to speak of magnitude of its elements. A *metric space* is a set (which need not be a vector space) endowed with a structure (a function called *metric* or *distance*) that allows us to speak of proximity of its elements. And the list of structures goes on. Of course, it can happen that we may have the same set endowed with distinct structures, say, two different inner products, so that we can have distinct spaces (two different inner product spaces) but the same set.

Definition 1.7 An **inner product** on a vector space V (over a field $\mathbb{F}$) is a real-valued function which associates a scalar (i.e., an element of the field) to each pair of vectors $(\mathbf{u}, \mathbf{v}) \in V \times V$, the scalar being denoted by $\mathbf{u} \cdot \mathbf{v}$, satisfying the following axioms: for all $\mathbf{u}, \mathbf{v}, \mathbf{w} \in V$ and $\lambda \in \mathbb{R}$,

I1. $\mathbf{u} \cdot \mathbf{u} \geq 0$ and $\mathbf{u} \cdot \mathbf{u} = 0$ if and only if $\mathbf{u} = \mathbf{0}$.
I2. $\mathbf{u} \cdot \mathbf{v} = \mathbf{v} \cdot \mathbf{u}$.
I3. $(\lambda \mathbf{u}) \cdot \mathbf{v} = \lambda(\mathbf{u} \cdot \mathbf{v})$.
I4. $(\mathbf{u} + \mathbf{v}) \cdot \mathbf{w} = \mathbf{u} \cdot \mathbf{w} + \mathbf{v} \cdot \mathbf{w}$.

A vector space together with an inner product is called an **inner product space** (or pre-Hilbert space). Another name for the inner product is *scalar product*, although usually this name is reserved for the case of finite-dimensional spaces.

There are many examples of inner product which are important in applications and we mention two of them that will be of interest for us. For the first one, consider the real vector space of all continuous functions from the closed interval $[a, b] \subset \mathbb{R}$ into $\mathbb{R}$ with the usual pointwise addition of functions and multiplication by a scalar. Then,

$$f \cdot g := \int_a^b f(t)g(t)dt, \text{ for all } f, g : [a, b] \to \mathbb{R} \text{ continuous},$$

defines an inner product on the vector space of all continuous real functions defined on $[a, b]$.

Another example will turn up fundamental for us in the sequel. It is the inner product defined on $\mathbb{R}^n$ by setting

$$\mathbf{u} \cdot \mathbf{v} = \sum_{i=1}^{n} u_i v^i = u_i v^i, \forall \mathbf{u} = (u_1, \ldots, u_n),\ \mathbf{v} = (v^1, \ldots, v^n) \in \mathbb{R}^n. \tag{1.3}$$

It can be shown that the operation defined by this formula satisfies the axioms I1–I4 above so that it defines an inner product on $\mathbb{R}^n$, called the *Euclidean inner product* or *dot product.* It is common to denote as E or E^n when referring to $\mathbb{R}^n$ as an *Euclidean inner product space.* The reason for the word Euclidean popping in will become clear after we introduce the notions of distance and norm.

Definition 1.8 Let M be a nonempty set. A **metric** (or **distance**) on M is a function, denoted $d(.,.)$, that associates to every pair (u, v) of elements of M the real number denoted $d(u, v)$ such that, for all $u, v, z \in M$, the following hold:

(D1) $d(u, v) \geq 0$ and $d(u, v) = 0$ if and only if $u = v$.
(D2) $d(u, v) = d(v, u)$.
(D3) $d(u, v) \leq d(u, z) + d(z, v)$.

Any pair (M, d), where M is a nonempty set and $d(.,.)$ is a metric defined on M, is called a **metric space**.

One important feature of metric spaces is that the notion of distance allows one to introduce the fundamental topological concepts of open balls and open sets that provide the framework to bring analytical tools such as convergence, limits, derivatives, and integrals into play.

Definition 1.9 Suppose (M, d) is a metric space and let $a \in M$ and $r > 0$ be given. We call **open ball** with center a and radius r, and denote $B(a, r)$, the set

$$B(a, r) = \{x \in M;\, d(x, a) < r\}.$$

A subset $S \subseteq M$ is said to be an **open set** if for each $x_o \in M$ there exists an open ball $B(x_o, r)$ such that $B(x_o, r) \subset S$.

Notice that the structure of metric space can be established on every nonempty set, regardless of being or not being imbued with the structure of vector space. On the other hand, the structures of inner product and the one that we shall introduce next are applicable only to vector spaces.

Definition 1.10 Let V be a vector space (over the field $\mathbb{F}$). A **norm** on V is a function, denoted $||\cdot||$, that assigns to each $\mathbf{u} \in V$ a real number, denoted $||\mathbf{u}||$, satisfying the following axioms:

(N1) $||\mathbf{u}|| \geq 0$ and $||\mathbf{u}|| = 0$ if and only if $\mathbf{u} = \mathbf{0}$.
(N2) $||\lambda \mathbf{u}|| = |\lambda| \cdot ||\mathbf{u}||$, for all $\lambda \in \mathbb{F}$.
(N3) $||\mathbf{u} + \mathbf{v}|| = ||\mathbf{u}|| + ||\mathbf{v}||$, for all $\mathbf{u}, \mathbf{v} \in V$.

We call **normed space** any vector space together with a norm defined on it. If $||\mathbf{u}|| = 1$, we say that $\mathbf{u}$ is a **unit vector**.

As an example, consider the vector space $\mathbb{R}^n$ over the field $\mathbb{R}$ with the usual operations. One can show that any of the following functions satisfy the axioms that define norm:

$$\begin{aligned} ||\mathbf{u}||_1 &= \sum_{i=1}^{n} |u_i| \\ ||\mathbf{u}||_2 &= \Big(\sum_{i=1}^{n} |u_i|^2\Big)^{1/2} \\ ||\mathbf{u}||_\infty &= \max\{|u_1|, \ldots, |u_n|\}. \end{aligned}$$

This example shows that the same vector space ($\mathbb{R}^n$ in this case) can have more than one norm defined on it, yielding different normed spaces.

Among the properties that can be deduced from the axioms defining the inner product, probably the most important one is the following result together with a corollary that establishes that every inner product space is a normed space with a norm induced by the inner product. The proofs can be found in any book on functional analysis.

Theorem 1.1 *(Cauchy–Schwarz inequality). If V is an inner product vector space, then*

$$|\mathbf{u} \cdot \mathbf{v}|^2 \leq (\mathbf{u} \cdot \mathbf{u})(\mathbf{v} \cdot \mathbf{v}), \text{ for all } \mathbf{u}, \mathbf{v} \in V, \tag{1.4}$$

where $|\mathbf{u} \cdot \mathbf{v}|$ denotes the absolute value of the real number $\mathbf{u} \cdot \mathbf{v}$. Equality occurs if and only if $\mathbf{u}$ and $\mathbf{v}$ are linearly dependent.

Corollary 1.1 *Let $\cdot$ be an inner product on a vector space V. Then,*

$$||\mathbf{u}|| = \sqrt{\mathbf{u} \cdot \mathbf{u}}, \ \forall \mathbf{u} \in V \tag{1.5}$$

defines a norm on V.

Furthermore, it can be shown that every normed space is naturally made into a metric space with the distance induced by the norm. More precisely, if V is a vector space imbued with a norm $||.||$, then one can show that

$$d(\mathbf{u}, \mathbf{v}) = ||\mathbf{u} - \mathbf{v}||, \ \forall \mathbf{u}, \mathbf{v} \in V,$$

defines a metric on V.

Putting together this sequence of naturally induced new structures that can be added on step by step on a vector space endowed with an inner product, we have that every inner product space V can be induced into a normed space by defining a norm as

$$||\mathbf{u} \cdot \mathbf{u}|| = \sqrt{\mathbf{u} \cdot \mathbf{u}}, \ \forall \mathbf{u} \in V,$$

and further induced into a metric space, by defining a distance as

$$d(\mathbf{u}, \mathbf{v}) = ||\mathbf{u} - \mathbf{v}|| = \sqrt{(\mathbf{u} - \mathbf{v}) \cdot (\mathbf{u} - \mathbf{v})}, \forall \mathbf{u}, \ \mathbf{v} \in V.$$

In some studies, it is important to know whether a particular norm on some vector space is induced by an inner product. It has been shown (see, for instance, Horn–Johnson [Horn and Johnson, 1985]) that the so-called parallelogram identity

$$\frac{1}{2}\left(||\mathbf{u}+\mathbf{v}||^2+||\mathbf{u}-\mathbf{v}||^2\right)=||\mathbf{u}||^2+||\mathbf{v}||^2$$

is a necessary and sufficient condition for a given norm to be derived from an inner product.

If $\mathbf{u},\mathbf{v}$ are any nonzero vectors in V, the Cauchy–Schwarz inequality (1.4) yields

$$|\mathbf{u}\cdot\mathbf{v}|\leq(\mathbf{u}\cdot\mathbf{u})^{1/2}(\mathbf{v}\cdot\mathbf{v})^{1/2},\ \forall\,\mathbf{u},\mathbf{v}\in V,$$

and hence

$$-1\leq\frac{\mathbf{u}\cdot\mathbf{v}}{(\mathbf{u}\cdot\mathbf{u})^{1/2}(\mathbf{v}\cdot\mathbf{v})^{1/2}}\leq 1,\ \forall\,\mathbf{u},\mathbf{v}\in V.$$

Thus, whenever a vector space is provided with an inner product, it makes sense to speak of angles between their vectors and orthogonality by defining an angle θ between any nonzero vectors $\mathbf{u}$ and $\mathbf{v}$ as

$$\theta=arc\cos\left(\frac{\mathbf{u}\cdot\mathbf{v}}{(\mathbf{u}\cdot\mathbf{u})^{1/2}(\mathbf{v}\cdot\mathbf{v})^{1/2}}\right).$$

Moreover, two vectors $\mathbf{u}$ and $\mathbf{v}$ are said to be *orthogonal* iff $\mathbf{u}\cdot\mathbf{v}=0$. In particular, a set of vectors in an inner product space is said to be *orthonormal* if all the vectors in this set have unit norm (the norm being the one induced by the inner product) and are mutually orthogonal.

Example 1.3 (a) Find the angle between the following vectors

$$\mathbf{u}=\begin{Bmatrix}1\\1\\0\end{Bmatrix}\text{ and }\mathbf{v}=\begin{Bmatrix}-1\\1\\1\end{Bmatrix},$$

on $\mathbb{R}^3$ with the Euclidean inner product.

(b) Show that any nonzero pairwise orthogonal vectors $\mathbf{v}_1,\mathbf{v}_2,\ldots,\mathbf{v}_n$ are linearly independent.

Solution:

(a) Since $\mathbf{u}\cdot\mathbf{v}=-1+1+0=0$, we have

$$\cos\theta=\frac{\mathbf{u}\cdot\mathbf{v}}{\sqrt{(\mathbf{u}\cdot\mathbf{u})(\mathbf{v}\cdot\mathbf{v})}}=0.$$

Thus, the angle is

$$\theta = \frac{\pi}{2}.$$

(b) Let $c_1, c_2, \ldots, c_n$ be n scalars such that $c_1\mathbf{v}_1 + c_2\mathbf{v}_2 + \cdots + c_n\mathbf{v}_n = \mathbf{0}$. Then,

$$c_1(\mathbf{v}_1 \cdot \mathbf{v}_1) + c_2(\mathbf{v}_2 \cdot \mathbf{v}_1) + \cdots + c_n(\mathbf{v}_n \cdot \mathbf{v}_1) = \mathbf{0}.$$

Now, from the orthogonality of the vectors it follows that this gives $c_1 = 0$, since $\mathbf{v}_1 \cdot \mathbf{v}_1 \neq \mathbf{0}$ and $\mathbf{v}_k \cdot \mathbf{v}_1 = \mathbf{0}$, for $k \neq 1$. Similar reasoning gives that all the other scalars are also zero.

Since in every inner product space we can meaningfully speak of angle and distance between its vectors, and this ultimately reduces to making sense to what we can draw having drawing compasses and rulers as tools in Geometry, we can say that an inner product structure provides the vector space with a Geometry in this sense. It happens that $\mathbb{R}^3$ provided with the particular inner product given by (see 1.3), that is,

$$\mathbf{u} \cdot \mathbf{v} = \sum_{i=1}^{3} u_i v^i \stackrel{\triangle}{=} u_i v^i, \quad \forall \mathbf{u} = (u_1, u_2, u_3), \ \mathbf{v} = (v^1, v^2, v^3) \in \mathbb{R}^3,$$

which we called Euclidean inner product by the way, turns up to have the metric

$$d(\mathbf{u}, \mathbf{v}) = ||\mathbf{u} - \mathbf{v}|| = \sqrt{\sum_{i=1}^{3} (u_i - v^i)^2}$$

measuring the standard Euclidean distance between two points in space (namely, the length of the straight line segment connecting the two points) that was established by Euclid in his *Elements*. With this remark in mind, we define:

Definition 1.11 We call **Euclidean space**, and denote by E or E^n, the real vector space $\mathbb{R}^n$ together with the inner product defined by

$$\mathbf{u} \cdot \mathbf{v} = \sum_{i=1}^{n} u_i v^i \stackrel{\triangle}{=} u_i v^i, \quad \forall \mathbf{u} = (u_1, \ldots, u_n), \ \mathbf{v} = (v^1, \ldots, v^n) \in \mathbb{R}^n,$$

and the norm induced by it, namely (dispensing the summation symbol $\sum_{k=1}^{n}$),

$$||\mathbf{u}|| = \sqrt{\mathbf{u}\cdot\mathbf{u}} = \sqrt{\sum_{i=1}^{n}(u_i u^i)} \triangleq \sqrt{(u_i u^i)}.$$

1.4 Contravariant and covariant components

We can express the structuring operations of inner product, norm, distance and angle on an Euclidean inner product space E in terms of the component representation of each vector with respect to a basis $\mathbf{g} = (\mathbf{g}_1, \ldots, \mathbf{g}_n)$ in E. For this, let g_{ij} denote the inner products $g_{ij} = \mathbf{g}_i \cdot \mathbf{g}_j$ of pairs of vectors from the basis. Then, given any pair of vectors $\mathbf{u} = u^i\mathbf{g}_i$ and $\mathbf{v} = v^i\mathbf{g}_i$ on E, their (Euclidean) inner product is given by

$$\mathbf{u}\cdot\mathbf{v} = u^i v^j \mathbf{g}_i \cdot \mathbf{g}_j = u^i v^j g_{ij}, \qquad i, j = 1, \ldots, n.$$

The norm of $\mathbf{u}$ is given by

$$||\mathbf{u}|| = \sqrt{u^i u^j g_{ij}}$$

and the angle θ between two nonzero vectors $\mathbf{u}$ and $\mathbf{v}$ is

$$\theta = \frac{u^i v^i g_{ij}}{\sqrt{u^i u^j g_{ij}}\sqrt{v^i v^j g_{ij}}}.$$

If a set $\{\mathbf{g}^1, \ldots, \mathbf{g}^n\} \subset E$ is such that

$$\mathbf{g}^i \cdot \mathbf{g}_j = \delta^i_j = \begin{cases} 1\,, & \text{if } i = j \\ 0\,, & \text{if } i \neq j \end{cases}, \qquad i, j = 1, \ldots, n,$$

we call it the *reciprocal set* of the basis $\{\mathbf{g}_1, \ldots, \mathbf{g}_n\}$. We have that $\mathbf{g}^i \cdot \mathbf{g}_j = \delta^i_j$ implies that $\{\mathbf{g}^1, \ldots, \mathbf{g}^n\}$ is linearly independent and therefore is also a basis of E. Thus, we have:

Definition 1.12 Two bases $\{\mathbf{g}_1, \ldots, \mathbf{g}_n\}$ and $\{\mathbf{g}^1, \ldots, \mathbf{g}^n\}$ in an Euclidean space E are said to be **reciprocal bases** iff

$$\mathbf{g}^i \cdot \mathbf{g}_j = \delta^i_j.$$

In this case, any $\mathbf{u} \in E$ can be represented as linear combinations of the elements of each basis, namely,

$$\mathbf{u} = u^i \mathbf{g}_i \quad \text{and} \quad \mathbf{u} = u_i \mathbf{g}^i, \tag{1.6}$$

and the scalars $u_1, \ldots, u_n$ are called **covariant components** of $\mathbf{u}$, and $u^1, \ldots, u^n$ are called **contravariant components** of $\mathbf{u}$.

We can express each one of these components in terms of the others as

$$u_k = u^i g_{ik} \quad \text{and} \quad u^k = g^{ik} u_i, \tag{1.7}$$

where $g_{ik} = \mathbf{g}_i \cdot \mathbf{g}_k$ and $g^{ik} = \mathbf{g}^i \cdot \mathbf{g}^k$. Indeed, applying the inner product by $\mathbf{g}_k$ to each side of (1.6), we have

$$u_i \mathbf{g}^i \cdot \mathbf{g}_k = u^i \mathbf{g}_i \cdot \mathbf{g}_k,$$

which gives $u_k = u^i g_{ik}$. Similarly, applying the inner product by $\mathbf{g}^k$ we end up with $u^k = g^{ik} u_i$.

Also, the inner product of arbitrary vectors $\mathbf{u}$ and $\mathbf{v}$ can be expressed in terms of reciprocal bases $\mathbf{g} = (\mathbf{g}_1, \ldots, \mathbf{g}_n)$ and $\mathbf{g}^* = (\mathbf{g}^1, \ldots, \mathbf{g}^n)$ in an Euclidean space $\mathbf{E}$. Indeed, taking

$$\mathbf{u} = u^i \mathbf{g}_i = u_i \mathbf{g}^i, \tag{1.8a}$$

$$\mathbf{v} = v^j \mathbf{g}_j = v_j \mathbf{g}^j, \tag{1.8b}$$

we have

$$\mathbf{u} \cdot \mathbf{v} = u^i v_i = u_i v^i = g_{ij} u^i v^j = g^{ij} u_i v_j.$$

Next, let us evaluate how the components of a vector $\mathbf{u}$ transform under a change of basis. So, let $\{\mathbf{g}_1, \ldots, \mathbf{g}_n\}$ and $\{\mathbf{g}'_1, \ldots, \mathbf{g}'_n\}$ be two bases in the Euclidean vector space $\mathbf{E}$. We can represent componentwise the elements of each one of them in terms of the elements of the other basis. Thus, we have

$$\mathbf{g}'_j = A^i_j \mathbf{g}_i, \tag{1.9}$$

$$\mathbf{g}_k = B^j_k \mathbf{g}'_j, \tag{1.10}$$

where each A^i_j and B^j_k are entries of nonsingular matrices. The components of each of these representations satisfy the relations

$$A^k_i B^j_k = \delta^j_i,$$

$$B^k_i A^j_k = \delta^j_i.$$

Indeed, substituting (1.9) into (1.10) we have $\mathbf{g}_k = B^j_k A^i_j \mathbf{g}_i$ and, since $\mathbf{g}_k = \delta^i_k \mathbf{g}_i$, we have $(B^j_k A^i_j - \delta^i_k)\mathbf{g}_i = 0$, so that $A^k_i B^j_k = \delta^j_i$. Substituting the other way around leads to $B^k_i A^j_k = \delta^j_i$.

In particular, if the bases are reciprocal, the relations $\mathbf{g}'^j = C^j_i \mathbf{g}^i$ and $\mathbf{g}^i = D^i_j \mathbf{g}'^j$, in conjunction with

$$\mathbf{u} = u_k \mathbf{g}^k = u^k \mathbf{g}_k = u'_j \mathbf{g}'^j = u'^j \mathbf{g}'_j,$$

yield that, under a change of bases, the components of a vector $\mathbf{u}$ transform according to

$$\begin{aligned} u'_j &= D^i_j u_i, \qquad u_j = C^i_j u'_i, \\ u'^j &= B^j_i u^i, \qquad u^j = A^j_i u'^i. \end{aligned} \tag{1.11}$$

1.5 Coordinate systems

Consider a nonempty set $\mathfrak{E}$, whose elements we denote by capital letters and call *points*, and an arbitrary reference point, which we call *origin* and denote O. Let E be a n-dimensional *Euclidean space.* We can establish a structure relating points in $\mathfrak{E}$ to vectors in E that allows us to introduce the concept of system of coordinates and formalize the analytical geometry.

Definition 1.13 We say that $\mathfrak{E}$ is a *Euclidean point space* iff for each ordered pair (X, Y) of points in $\mathfrak{E}$ there corresponds a unique vector in E, which we denote $\overrightarrow{\mathrm{XY}}$, such that

(E1) $\overrightarrow{\mathrm{PQ}} = -\overrightarrow{\mathrm{QP}}$.

(E2) $\overrightarrow{\mathrm{XY}} = \overrightarrow{XZ} + \overrightarrow{ZY}$, for any $\mathrm{Z} \in \mathfrak{E}$.

(E3) To each vector $\mathbf{x} \in E$ there corresponds a unique point $\mathrm{X} \in \mathfrak{E}$ such that $\mathbf{x} = \overrightarrow{\mathrm{OX}}$.

It is convenient to write $\overrightarrow{\mathrm{XY}}$ in the form

$$\overrightarrow{\mathrm{XY}} = \mathbf{y} - \mathbf{x} := \mathbf{v}.$$

Definition 1.14 $\mathfrak{E}$ and E constitute an **affine space** iff there exists a function s, called sum of point and vector, that associates to each pair $(\mathrm{P}, \mathbf{u})$ in $\mathfrak{E} \times E$ a point $s(\mathrm{P}, \mathbf{u})$ in $\mathfrak{E}$, which we denote $s(\mathrm{P}, \mathbf{u}) = \mathbf{P} + \mathbf{u}$, such that[2]

(A1) $\forall\ \mathrm{P} \in \mathfrak{E}$ and $\mathbf{u}, \mathbf{v} \in E$ we have

$$\mathbf{P} + (\mathbf{u} + \mathbf{v}) = (\mathbf{P} + \mathbf{u}) + \mathbf{v}.$$

(A2) For all $(\mathrm{P}, \mathrm{Q}) \in \mathfrak{E} \times \mathfrak{E}$ there exists a unique $\mathbf{u} \in E$ such that $\mathbf{Q} = \mathbf{P} + \mathbf{u}$.

---

[2]Note that P is the name of a point, whereas $\mathbf{P}$ is the vector $\overrightarrow{\mathrm{OP}}$.

As a straightforward consequence of the definition of affine space we have that, $\forall \mathrm{P} \in \mathfrak{E}$,

$$\mathbf{P} + \mathbf{0} = \mathbf{P},$$

where $\mathbf{0}$ is the zero vector in E. It is convenient to denote the unique $\mathbf{u}$ of the definition as $\mathbf{u} = \overrightarrow{\mathrm{PQ}}$ and to write $\overrightarrow{\mathrm{XY}}$ in the form

$$\overrightarrow{\mathrm{XY}} = \mathbf{u} = \mathbf{y} - \mathbf{x}$$

so that the distance between the points X and Y in $\mathfrak{E}$, denoted $d(\mathbf{x}, \mathbf{y})$, is given by the norm of $\mathbf{u}$:

$$d(\mathbf{x}, \mathbf{y}) = ||\mathbf{u}|| = \sqrt{(\mathbf{y} - \mathbf{x}) \cdot (\mathbf{y} - \mathbf{x})}.$$

We define the angle θ between the lines joining the points X and Y to an arbitrary origin O in $\mathfrak{E}$ by

$$\theta = \arccos\left(\frac{\mathbf{x} \cdot \mathbf{y}}{||\mathbf{x}||.||\mathbf{y}||}\right).$$

In the study of regions (or bodies) of more general vector spaces, Élie Cartan and others introduced and developed extensively the notion of *frames* or marks. In this setting, a frame carries the geometric idea of a basis of a vector space. Some examples of frames are [Cotton, 1905]:

(1) A *linear frame* is an ordered basis of a vector space.
(2) An *affine frame* of a vector space V consists of a choice of origin for V along with an ordered basis of vectors in V.
(3) An *orthonormal frame* of a vector space is an ordered basis consisting of orthogonal unit vectors (an orthonormal basis).
(4) A *Euclidean frame* of a vector space is a choice of origin along with orthonormal basis for the vector space.
(5) A *Darboux frame* is any of the moving frames defined above that is moving on a surface in an Euclidean space instead of a curve.

Definition 1.15
(i) A **rectangular Cartesian coordinate system** is a Euclidean frame in the Euclidean space $E = \mathbb{R}^n$ consisting of an origin $\mathrm{O} \in \mathfrak{E} = \mathbb{R}^n$ together with an orthonormal basis $\{\mathbf{i}_1, \mathbf{i}_2, \ldots, \mathbf{i}_n\}$ in $\mathbb{R}^n$. Thus, to every point $\mathrm{X} \in \mathfrak{E} = \mathbb{R}^n$ there corresponds a unique vector $\mathbf{x} = \overrightarrow{\mathrm{OX}} \in E = \mathbb{R}^n$ uniquely represented as

$$\mathbf{x} = x^j \mathbf{i}_j, \text{with} \quad x^j = \mathbf{x} \cdot \mathbf{i}_j.$$

(ii) The scalars $x^1, x^2, \ldots, x^n$ above, or the n-tuple $(x^1, x^2, \ldots, x^n)$, are called **coordinates of the point** X relative to the rectangular Cartesian coordinate system $\{\mathrm{O}, \mathbf{i}_1, \mathbf{i}_2, \ldots, \mathbf{i}_n\}$.
(iii) We define the **position vector** of a point P in the space $\mathbb{R}^n$ to be the vector

$$\mathbf{r} = \overrightarrow{\mathrm{OP}} := \mathbf{P} = x^j \mathbf{i}_j.$$

Although this definition concerns n-dimensional vector spaces in general, for the purposes of this book we consider from now on the case $n = 3$ to correspond to the tridimensional geometric setup that serves as the background for the study of elastic deformations on rigid bodies. Figure 1.1 illustrates tridimensional Cartesian frames $\mathcal{F}_k$ with origins O_k, $\mathcal{F}_k = \{\mathrm{O}_k, \mathbf{i}_1, \mathbf{i}_2, \mathbf{i}_3\}$, for $k = 1, 2, 3$.

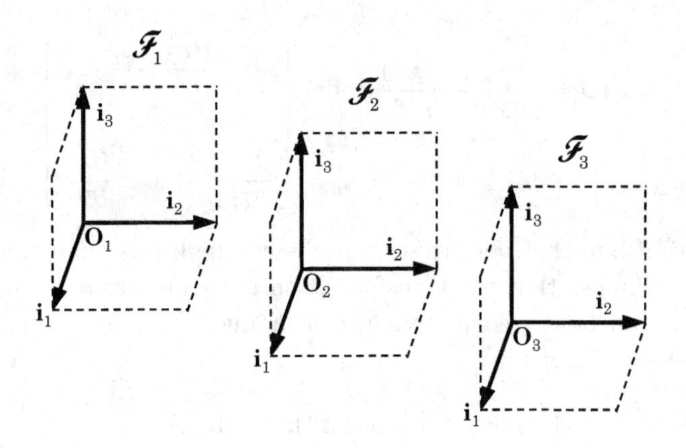

Fig. 1.1 Three tridimensional Cartesian frames.

1.6 Change of coordinate systems

We remind the reader that, for a given nonempty open set $\Omega \subset \mathbb{R}^n$ and an integer $k \geq 1$, we say that a function $f : \Omega \to \mathbb{R}$ is of *class* C^k on Ω, denoted $f \in C^k(\Omega)$, if there exist and are continuous in Ω the partial derivatives

$$\frac{\partial^k f}{\partial x_1^{i_1} \partial x_2^{i_2} \cdots \partial x_n^{i_n}},$$

for all nonnegative integers $i_1, i_2, \ldots, i_n$ such that $i_1 + i_2 + \cdots + i_n = k$. A function of class C^1 sometimes is said to be *continuously differentiable.*

Let Ω be a given open subset of $\mathfrak{E} = \mathbb{R}^n$. We may relate this set to a Cartesian system of axes determined by a Euclidean frame $\{\mathbf{i}_1, \ldots, \mathbf{i}_n\}$ in the following way. To each point X in Ω, we take the representation of its position vector $\mathbf{x} \in E$ in terms of this basis, namely, $\mathbf{x} = \overline{x}^k \mathbf{i}_k$, where $(\overline{x}^1, \ldots, \overline{x}^n)$ are the Cartesian coordinates of the point X in Ω (formally, we also say $\mathbf{x} \in \Omega$). A new set of general coordinates can be introduced by means of the transformation

$$x^i = x^i(\overline{x}^1, \ldots, \overline{x}^n), \quad i = 1, 2, \cdots, n, \tag{1.12}$$

with x^i representing single-valued, continuously differentiable functions defined at all points $(\overline{x}^1, \ldots, \overline{x}^n) \in \Omega$ that are independent of each other in the sense that the Jacobian determinant

$$\det J = \frac{\partial(x^1, \ldots, x^n)}{\partial(\overline{x}^1, \ldots, \overline{x}^n)} = \det \begin{bmatrix} \dfrac{\partial x^1}{\partial \overline{x}^1} & \cdots & \cdots & \dfrac{\partial x^1}{\partial \overline{x}^n} \\ \vdots & \dfrac{\partial x^2}{\partial \overline{x}^2} & \cdots & \vdots \\ \vdots & \vdots & \ddots & \vdots \\ \dfrac{\partial x^n}{\partial \overline{x}^1} & \cdots & \cdots & \dfrac{\partial x^n}{\partial \overline{x}^n} \end{bmatrix}$$

does not vanish in Ω. Under these hypotheses, the Inverse Function Theorem of Calculus implies that the transformation is locally invertible in the sense that (1.12) can be solved in terms of a unique set of inverse continuously differentiable functions

$$\overline{x}^i = \overline{x}^i(x^1, x^2, \cdots, x^n), \; i = 1, 2, \cdots, n \tag{1.13}$$

on some open subset $S \subset \Omega$.

The continuously differentiable transformations (1.12) and (1.13) establish a one-to-one correspondence between the Cartesian coordinates $(\overline{x}^1, \ldots, \overline{x}^n)$ of the point $\mathbf{x} \in \Omega$ and the variables $(x^1, x^2, \cdots, x^n) \in S$. We say that the space E has been referred on Ω to a *system of curvilinear coordinates* $(x^1, x^2, \cdots, x^n)$.

For the functions x^i introduced in (1.12) and arbitrary real constants $c^1, \ldots, c^n$, the equations

$$\left.\begin{aligned} x^1(\overline{x}^1, \ldots, \overline{x}^n) &= c^1, \\ x^2(\overline{x}^1, \ldots, \overline{x}^n) &= c^2, \\ &\vdots \\ x^n(\overline{x}^1, \ldots, \overline{x}^n) &= c^n, \end{aligned}\right. \qquad \text{for all } \mathbf{x} = (\overline{x}^1, \ldots, \overline{x}^n) \in \Omega,$$

define surfaces $S_1, S_2, \ldots, S_n$ on $\mathbb{R}^n$. Each surface obtained by setting

$$x^i(\overline{x}^1, \ldots, \overline{x}^n) = c^i, \quad \forall (\overline{x}^1, \ldots, \overline{x}^n) \in \Omega,$$

is called the x^i-*coordinate surface*. The intersections of $n-1$ of these coordinate surfaces are called the coordinate lines $x^1, x^2, \cdots, x^n$. These lines intersect at a point.

As an example, in the case $n = 3$ the x^1-coordinate line is the intersection of the surfaces $x^2 = c^2$ and $x^3 = c^3$. Along this line, the only variable that undergoes any change is x^1.

Consider the special type of curvilinear coordinate system known as a *spherical coordinate system* with a pole located at the origin O of a given Cartesian system Oxyz (see Figure 1.2). The spherical coordinates

$$x^1 = r, \quad x^2 = \theta, \quad x^3 = \varphi$$

are related to the rectangular coordinates

$$\overline{x}^1 = x, \quad \overline{x}^2 = y, \quad \overline{x}^3 = z$$

by means of the transformation

$$x = r \sin\theta \cos\varphi, \quad y = r \sin\theta \sin\varphi, \quad z = r\cos\theta. \tag{1.14}$$

The inverse of (1.14) is

$$r = \sqrt{x^2 + y^2 + z^2}, \quad \theta = \tan^{-1}\frac{\sqrt{x^2+y^2}}{z}, \quad \varphi = \tan^{-1}\frac{y}{x},$$

which is single-valued whenever $r > 0$, $0 < \theta < \pi$, and $0 \le \varphi < 2\pi$.

The coordinate surfaces $r =$ constant, $\theta =$ constant, and $\varphi =$ constant are spheres, cones, and semiplanes, respectively. The coordinate lines are the meridians (intersection of $r =$ constant and $\varphi =$ constant), the parallel lines (intersection of $\theta =$ constant and $r =$ constant), and the radial lines (intersection of $\varphi =$ constant and $\theta =$ constant).

In the case of the *cylindrical coordinate system*, the coordinates (r, θ, z) are related to the rectangular ones (x, y, z) by means of the transformations (see Figure 1.3)

$$x = r\cos\theta, \quad y = r\sin\theta, \quad z = z.$$

The inverse transformation,

$$r = \sqrt{x^2 + y^2}, \quad \theta = \tan^{-1}\frac{y}{x}, \quad z = z,$$

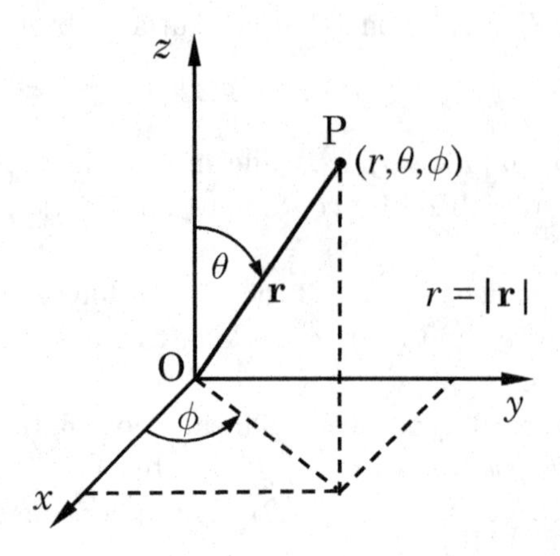

Fig. 1.2 Spherical coordinate system.

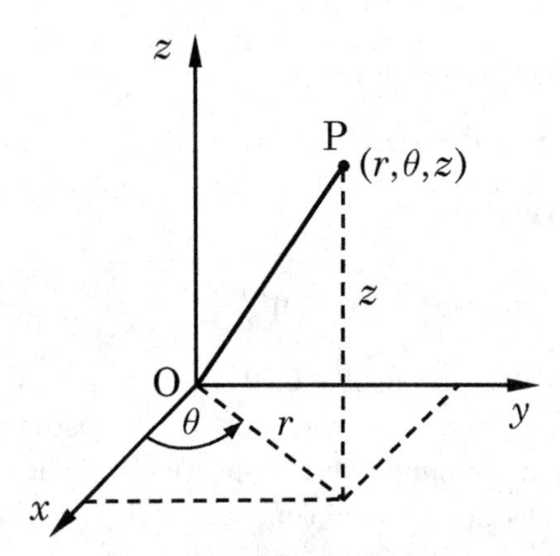

Fig. 1.3 Cylindrical coordinate system.

is single-valued whenever $0 \leq \theta \leq 2\pi$ and $r > 0$. The coordinate surfaces $r =$ constant, $\theta =$ constant, and $z =$ constant are circular cylinders, planes containing the z axis, and planes perpendicular to the z axis, respectively. The coordinate lines are the generators of the cylindrical surfaces (intersection of $r =$ constant and $\theta =$ constant). The parallel lines (intersection of $r =$ constant and $z =$ constant) and the radial lines (intersection of $\theta =$ constant and $z =$ constant surfaces) .

1.7 Exercises

(1) Consider $S = \{\mathbf{v}_1, \mathbf{v}_2, \mathbf{v}_3\}$, where

$$\mathbf{v}_1 = \begin{Bmatrix} 1 \\ 1 \\ -1 \end{Bmatrix}, \quad \mathbf{v}_2 = \begin{Bmatrix} 1 \\ 0 \\ -1 \end{Bmatrix}, \quad \mathbf{v}_3 = \begin{Bmatrix} 0 \\ -1 \\ 0 \end{Bmatrix}.$$

Find a basis for *spanS*. *Hint*: Note that $\{\mathbf{v}_1, \mathbf{v}_2\}$ is linearly independent.

(2) Find the dimension of

$$span \left\{ \begin{Bmatrix} -2 \\ -1 \\ 1 \\ 0 \end{Bmatrix}, \begin{Bmatrix} -1 \\ 0 \\ 1 \\ 1 \end{Bmatrix}, \begin{Bmatrix} 2 \\ 2 \\ 0 \\ 2 \end{Bmatrix}, \begin{Bmatrix} 0 \\ 1 \\ 1 \\ 2 \end{Bmatrix}, \begin{Bmatrix} 1 \\ 1 \\ 0 \\ 1 \end{Bmatrix} \right\}.$$

Answer: 2.

(3) Show that the set of all vectors of the form

$$\begin{Bmatrix} a \\ b \\ 0 \end{Bmatrix}, \quad a, b \in \mathbb{R},$$

with the usual addition and multiplication by a scalar, is a vector space over $\mathbb{R}$. *Hint*: Show that any vector in the set lies on the span of

$$\left\{ \begin{Bmatrix} 1 \\ 0 \\ 0 \end{Bmatrix}, \begin{Bmatrix} 0 \\ 1 \\ 0 \end{Bmatrix} \right\}.$$

(4) Let $\mathbf{M}$ be the vector space of all 2×2 matrices with real entries. Show that the following set forms a basis for $\mathbf{M}$:

$$\mathbf{B} = \left\{ \begin{bmatrix} 1 & 0 \\ 0 & 0 \end{bmatrix}, \begin{bmatrix} 0 & 1 \\ 0 & 0 \end{bmatrix}, \begin{bmatrix} 0 & 0 \\ 1 & 0 \end{bmatrix}, \begin{bmatrix} 0 & 0 \\ 0 & 1 \end{bmatrix} \right\}.$$

(5) (a) Given the basis $\mathbf{B} = \{g_1, g_2, g_3\}$, where

$$g_1 = \begin{Bmatrix} 1 \\ 1 \\ 1 \end{Bmatrix}, \quad g_2 = \begin{Bmatrix} 2 \\ 3 \\ 2 \end{Bmatrix}, \quad g_3 = \begin{Bmatrix} 3 \\ 3 \\ 4 \end{Bmatrix},$$

find its reciprocal basis.
(b) If $\mathbf{v} = 3g_1 + g_2 + 2g_3$, find the covariant components of $\mathbf{v}$.

(c) Evaluate $g_{ik} = \mathbf{g}_i \cdot \mathbf{g}_k$.
Answer:

$$\text{(a)} \qquad g^1 = \begin{Bmatrix} 6 \\ -1 \\ 3 \end{Bmatrix}, \; g^2 = \begin{Bmatrix} -1 \\ 1 \\ 0 \end{Bmatrix}, \; g^3 = \begin{Bmatrix} -1 \\ 0 \\ 1 \end{Bmatrix}.$$

(6) Given

$$\mathbf{v} = \begin{Bmatrix} 3 \\ 0 \\ 4 \end{Bmatrix}, \qquad \mathbf{w} = \begin{Bmatrix} -1 \\ 1 \\ 1 \end{Bmatrix},$$

find:
(a) $\mathbf{v} \cdot \mathbf{w}$,
(b) $|\mathbf{v}|$ and $|\mathbf{w}|$,
(c) $|\mathbf{v} + \mathbf{w}|$,
(d) $|\mathbf{v}| + |\mathbf{w}|$.
Answers: (a) 1, (b) 5 and $\sqrt{3}$, (c) $\sqrt{30}$, (d) $5 + \sqrt{3}$.

(7) Let $\{g_1 = \mathbf{i},\; g_2 = \mathbf{j}\}$ and

$$\left\{ \overline{g}_1 = \begin{Bmatrix} \cos\alpha \\ -\sin\alpha \end{Bmatrix}, \; \overline{g}_2 = \begin{Bmatrix} \sin\alpha \\ \cos\alpha \end{Bmatrix} \right\}$$

be two bases in $\mathbb{R}^2$. Show that the angle between $\mathbf{i}$ and $\overline{g}_1$ is α and the angle between $\mathbf{j}$ and $\overline{g}_2$ is also α.

(8) Show that the following set of vectors forms an orthogonal basis for $\mathbf{V} = \mathbb{R}^3$:

$$\left\{ \overline{g}_1 = \begin{Bmatrix} 1 \\ 0 \\ 1 \end{Bmatrix}, \; \overline{g}_2 = \begin{Bmatrix} -1 \\ 0 \\ 1 \end{Bmatrix}, \; \overline{g}_3 = \begin{Bmatrix} 0 \\ 312 \\ 0 \end{Bmatrix} \right\}.$$

(9) Suppose that the transformation

$$\begin{aligned} \overline{x} &= x \cdot x, \\ \overline{y} &= y \cdot y. \end{aligned}$$

connects the coordinate systems $(x^i) = (x, y)$ and $(\overline{x}^i) = (\bar{x}, \bar{y})$.
(a) Compute the Jacobian matrix of the transformation as well as its inverse.
(b) Calculate the vectors $\overline{\mathbf{g}}_1$ and $\overline{\mathbf{g}}_2$.

Answers:

$$\text{(a)}\ J = \begin{bmatrix} 2x & 0 \\ 0 & 2y \end{bmatrix}, \quad J^{-1} = \frac{1}{2}\begin{bmatrix} \frac{1}{x} & 0 \\ 0 & \frac{1}{y} \end{bmatrix}.$$

$$\text{(b)}\ \overline{\mathbf{g}}_1 = \frac{1}{2x}\mathbf{i}, \quad \overline{\mathbf{g}}_2 = \frac{1}{2y}\mathbf{j}.$$

(10) The spherical coordinates (x^i) are connected to rectangular coordinates $(\overline{x}^i)$ via

$$\begin{aligned} \overline{x}^1 &= x^1 \sin x^2 \cos x^3, \\ \overline{x}^2 &= x^1 \sin x^2 \sin x^3, \\ \overline{x}^3 &= x^1 \cos x^2. \end{aligned}$$

Find:
(a) The natural basis of the spherical coordinates.
(b) The Jacobian matrix

$$J = \left[\frac{\partial \overline{x}}{\partial x}\right]_{3\times 3}.$$

(c) The metric coefficients in matrix form, i.e., $G = J^T J$.

Answers:

(a)

$$\begin{aligned} \mathbf{g}_1 &= \sin x^2 \cos x^3 \mathbf{i} + \sin x^2 \sin x^3 \mathbf{j} + \cos x^2 \mathbf{k}, \\ \mathbf{g}_2 &= x^1(\cos x^2 \cos x^3 \mathbf{i} + \cos x^2 \sin x^3 \mathbf{j} - \sin x^2 \mathbf{k}), \\ \mathbf{g}_3 &= x^1 \sin x^2(-\sin x^3 \mathbf{i} + \cos x^3 \mathbf{j}). \end{aligned}$$

(b)

$$J = \begin{bmatrix} \sin x^2 \cos x^3 & x^1 \cos x^2 \cos x^3 & -x^1 \sin x^2 \sin x^3 \\ \sin x^2 \sin x^3 & x^1 \cos x^2 \sin x^3 & -x^1 \sin x^2 \cos x^3 \\ \cos x^2 & -x^1 \sin x^2 & 0 \end{bmatrix}.$$

(c)

$$G = J^T J = \begin{bmatrix} 1 & 0 & 0 \\ 0 & (x^1)^2 & 0 \\ 0 & 0 & (x^1 \sin x^2)^2 \end{bmatrix},$$

that is, $g_{11} = 1,\ g_{22} = (x^1)^2$, and $g_{33} = (x^1 \sin x^2)^2$.

(11) The parabolic coordinates (u, v, ϕ) are connected to rectangular orthonormal Cartesian coordinates $(\bar{x}^i) = (x, y, z)$ by

$$
\begin{aligned}
x &= uv \cos\phi, \\
y &= uv \sin\phi, \\
z &= \tfrac{1}{2}(v^2 - u^2).
\end{aligned}
$$

Find the metric coefficients g_{ij}.
Answer: $g_{ij} = 0$, if $i \neq j$, $g_{11} = g_{22} = (u^2 + v^2)^2$, and $g_{33} = (u^2 v^2)^2$.

(12) Calculate the volume element dV corresponding to the parabolic coordinates (u, v, ϕ) given by

$$
\begin{aligned}
x &= uv \cos\phi, \\
y &= uv \sin\phi, \\
z &= \tfrac{1}{2}(v^2 - u^2).
\end{aligned}
$$

Answer: $dV = (uv\sqrt{u^2 + v^2}) du dv d\phi$.

(13) The toroidal coordinates (r, θ, ϕ) are connected to rectangular orthonormal Cartesian coordinates $(\bar{x}^i) = (x, y, z)$ by

$$
\begin{aligned}
x &= (a - r\cos\theta)\cos\phi, \\
y &= (a - r\cos\theta)\sin\phi, \\
z &= r\sin\phi.
\end{aligned}
$$

where $a > 0$ is a given parameter. Find the metric coefficients g_{ij}.
Answer: $g_{ij} = 0$, if $i \neq j$, $g_{11} = 1$, $g_{22} = r^2$, and $g_{33} = (a - r\cos\theta)^2$.

Chapter 2

Vector and Tensor Algebras

The vector product, also called cross product, is an operation defined on the three-dimensional Euclidean space that associates to each pair of vectors a third vector in a prescribed manner inspired on both geometrical and physical applications, for instance, the velocity vector of a point within a solid body rotating steadily about an axis and the angular momentum about the origin of a moving particle. In Geometry, the equation of a straight line can be written in terms of a vector product.

2.1 Vector algebra

Definition 2.1 Let $\mathbf{E}$ be a three-dimensional Euclidean vector space. The vector product of two vectors $\mathbf{u}$ and $\mathbf{v}$ in $\mathbf{E}$ is a vector, denoted $\mathbf{u} \times \mathbf{v}$, whose direction is perpendicular to both $\mathbf{u}$ and $\mathbf{v}$ in a right-handed sense and whose magnitude is given by $|\mathbf{u}|.|\mathbf{v}| \ \sin\theta$, where θ is the angle between $\mathbf{u}$ and $\mathbf{v}$.

Thus,

$$\mathbf{u} \times \mathbf{v} = (|\mathbf{u}|\,|\mathbf{v}|\,\sin\theta)\,\mathbf{e} \tag{2.1}$$

where $\mathbf{e}$ is a unit vector perpendicular to $\mathbf{u}$ and $\mathbf{v}$ in a right-handed sense. By "perpendicular to both $\mathbf{u}$ and $\mathbf{v}$ in a right-handed sense" we mean that a right-handed screw rotated from $\mathbf{u}$ towards $\mathbf{v}$ moves in the direction of $\mathbf{u} \times \mathbf{v}$.

In order to have the vector product uniquely defined for each pair of vectors we need to introduce the notion of orientation for the vector space. To do this, consider the set $\mathcal{B}$ of all basis of a three-dimensional Euclidean vector space $\mathbf{E}$. Given any two bases for $\mathbf{E}$, say, $(\mathbf{g}) = (\mathbf{g}_1, \mathbf{g}_2, \mathbf{g}_3)$ and $(\mathbf{h}) = (\mathbf{h}_1, \mathbf{h}_2, \mathbf{h}_3)$, the nonsingular matrix $[a_{ij}]$ defined by

$$\mathbf{h}_j = \sum_{i=1}^{3} \mathbf{a}_{ij}\mathbf{g}_i, \quad j = 1, 2, 3,$$

is called the *matrix of change of the basis* (g) to the basis (h). We say that $(\mathbf{g})$ has the *same orientation* as $(\mathbf{h})$, and write

$$(\mathbf{g})\ eqv\ (\mathbf{h})$$

if and only if $\det[a_{ij}] > 0$. One can easily show that eqv is an equivalence relation and as such it provides a partition on $\mathcal{B}$ into two equivalence classes. More precisely, let the ordered triple $(\mathbf{e}_1, \mathbf{e}_3, \mathbf{e}_2)$ be a given basis in $\mathcal{B}$ and take the classes

$$\mathcal{B}_1 = \{(\mathbf{x}_1, \mathbf{x}_2, \mathbf{x}_3) \in \mathcal{B}; (\mathbf{x}_1, \mathbf{x}_2, \mathbf{x}_3)\ eqv\ (\mathbf{e}_1, \mathbf{e}_2, \mathbf{e}_3)\},$$
$$\mathcal{B}_2 = \{(\mathbf{x}_1, \mathbf{x}_2, \mathbf{x}_3) \in \mathcal{B}; (\mathbf{x}_1, \mathbf{x}_2, \mathbf{x}_3)\ eqv\ (\mathbf{e}_1, \mathbf{e}_3, \mathbf{e}_2)\}.$$

We have $\mathcal{B}_1 \cap \mathcal{B}_2 = \varnothing$ and $\mathcal{B}_1 \cup \mathcal{B}_2 = \mathcal{B}$. Each one of these classes $\mathcal{B}_1$ and $\mathcal{B}_2$ is called an *orientation* for $\mathbf{E}$ and we say that $\mathbf{E}$ is an *oriented Euclidean vector space* if and only if we fix one orientation for $\mathbf{E}$. In this case, the bases of the chosen orientation are called *positive bases.* Otherwise they are the *negative bases.*

It is usual to give an orientation to $\mathbf{E} = \mathbb{R}^3$ by considering $\mathcal{B}_1$ to be the orientation that contains the basis $(\mathbf{i}, \mathbf{j}, \mathbf{k})$ with $\mathbf{i} = (1, 0, 0)$, $\mathbf{j} = (0, 1, 0)$, and $\mathbf{k} = (0, 0, 1)$. This orientation is known as the *canonical orientation of* $\mathbb{R}^3$.

Definition 2.2 Let $(\mathbf{E}, \cdot, \mathcal{B}_1)$ be a three-dimensional oriented Euclidean vector space, where $\cdot$ stands for the scalar product. The *vector product* of two vectors $\mathbf{u}$ and $\mathbf{v}$ is a vector denoted by $\mathbf{u} \times \mathbf{v}$, such that

1. if $(\mathbf{u}, \mathbf{v})$ is linearly independent, then
   (a) $(\mathbf{u} \times \mathbf{v}) \cdot \mathbf{u} = (\mathbf{u} \times \mathbf{v}) \cdot \mathbf{v} = \mathbf{0}$ (i.e., $\mathbf{u} \times \mathbf{v}$ is perpendicular to both $\mathbf{u}$ and $\mathbf{v}$),
   (b) $|\mathbf{u} \times \mathbf{v}| = |\mathbf{u}|.|\mathbf{v}| \sin \boldsymbol{\theta}$, where θ is the angle between $\mathbf{u}$ and $\mathbf{v}$, and
   (c) $(\mathbf{u}, \mathbf{v}, \mathbf{u} \times \mathbf{v}) \in \mathcal{B}_1$.
2. if $(\mathbf{u}, \mathbf{v})$ is linearly dependent, then $\mathbf{u} \times \mathbf{v} = \mathbf{0}$.

If $\mathbf{E} = \mathbb{R}^3$ with the canonical orientation, we can write

$$\mathbf{u} \times \mathbf{v} = |\mathbf{u}|.|\mathbf{v}| \sin \boldsymbol{\theta}\ \mathbf{e},$$

where $\mathbf{e}$ is a unit vector perpendicular to $\mathbf{u}$ and $\mathbf{v}$ in the right-handed sense. Physically speaking, by this we mean that a right-handed screw rotated from $\mathbf{u}$ towards $\mathbf{v}$ moves in the direction of $\mathbf{u} \times \mathbf{v}$.

It follows from the definition that the vector product has some interesting properties:

(V1) Because of the right-hand rule, we have that

$$\mathbf{u} \times \mathbf{v} = -\mathbf{v} \times \mathbf{u}$$

so that the cross product is not commutative.

(V2) If $\mathbf{u}$ and $\mathbf{v}$ are parallel, then

$$\mathbf{u} \times \mathbf{v} = \mathbf{0}.$$

In particular, $\mathbf{u} \times \mathbf{u} = \mathbf{0}$.

(V3) $(\lambda\mathbf{u}) \times \mathbf{v} = \mathbf{u} \times (\lambda\mathbf{v}) = \lambda(\mathbf{u} \times \mathbf{v})$.

Since the area of the parallelogram made by the vectors $\mathbf{u}$ and $\mathbf{v}$ is the length of its base, say $|\mathbf{u}|$, multiplied by its height $|\mathbf{v}|.\sin\theta$, we have that $|\mathbf{u} \times \mathbf{v}|$ is the area of the parallelogram made by the two vectors. Similarly, the area of the triangle made by $\mathbf{u}$ and $\mathbf{v}$ is $\dfrac{|\mathbf{u} \times \mathbf{v}|}{2}$.

It can be shown geometrically that the vector product is distributive over addition:

$$\mathbf{u} \times (\mathbf{v} + \mathbf{w}) = \mathbf{u} \times \mathbf{v} + \mathbf{u} \times \mathbf{w}.$$

Also, we can express the vector product $\mathbf{u} \times \mathbf{v}$ in terms of the components of $\mathbf{u}$ and $\mathbf{v}$ by means of the following procedure. Let $(\mathbf{i}, \mathbf{j}, \mathbf{k})$ be an orthonormal basis of $\mathbf{E}$, ordered in a right-handed sense so that the unit vectors $\mathbf{i}, \mathbf{j}$, and $\mathbf{k}$ obey the following rules for their mutual cross product according to the definition above:

$$\mathbf{i} \times \mathbf{i} = \mathbf{j} \times \mathbf{j} = \mathbf{k} \times \mathbf{k} = \mathbf{0},$$
$$\mathbf{i} \times \mathbf{j} = \mathbf{k}, \quad \mathbf{j} \times \mathbf{k} = \mathbf{i}, \quad \mathbf{k} \times \mathbf{i} = \mathbf{j}.$$

Let

$$\mathbf{u} = u_1\mathbf{i} + u_2\mathbf{j} + u_3\mathbf{k},$$
$$\mathbf{v} = v_1\mathbf{i} + v_2\mathbf{j} + v_3\mathbf{k}$$

be the representation of $\mathbf{u}$ and $\mathbf{v}$ in terms of the basis. Then,

$$\begin{aligned}\mathbf{u} \times \mathbf{v} =& (u_1\mathbf{i} + u_2\mathbf{j} + u_3\mathbf{k}) \times (v_1\mathbf{i} + v_2\mathbf{j} + v_3\mathbf{k})\\ =& u_1v_1\mathbf{i} \times \mathbf{j} + u_1v_3\mathbf{i} \times \mathbf{k} + u_2v_1\mathbf{j} \times \mathbf{i} +\\ &+ u_2v_3\mathbf{j} \times \mathbf{k} + u_3v_1\mathbf{k} \times \mathbf{i} + u_3v_2\mathbf{k} \times \mathbf{j}\\ =& (u_2v_3 - u_3v_2)\mathbf{i} + (u_3v_1 - u_1v_3)\mathbf{j} + (u_1v_2 - u_2v_1)\mathbf{k}.\end{aligned}$$

This can be written in an abbreviated manner as the formal determinant

$$\mathbf{u} \times \mathbf{v} = \det \begin{bmatrix} \mathbf{i} & \mathbf{j} & \mathbf{k} \\ u_1 & u_2 & u_3 \\ v_1 & v_2 & v_3 \end{bmatrix}.$$

It follows from (2.1) by direct computation that

$$\mathbf{u} \times (\mathbf{v} \times \mathbf{w}) = (\mathbf{u} \cdot \mathbf{w})\mathbf{v} - (\mathbf{u} \cdot \mathbf{v})\mathbf{w}.$$

Geometrically, any three linearly independent vectors $\mathbf{u}, \mathbf{v}, \mathbf{w} \in \mathbf{E}$ form the three-dimensional geometric object known as a parallelepiped. The base of the parallelepiped is the parallelogram with area given by $|\mathbf{v} \times \mathbf{w}|$ and its height is the magnitude of the component of $\mathbf{u}$ in the direction of $\mathbf{v} \times \mathbf{w}$, which is

$$\frac{|\mathbf{u} \cdot (\mathbf{v} \times \mathbf{w})|}{|\mathbf{v} \times \mathbf{w}|}.$$

Therefore, the volume of the parallelepiped formed by $\mathbf{u}, \mathbf{v}$, and $\mathbf{w}$, which is the area of the base multiplied by the height, is $|\mathbf{u} \cdot (\mathbf{v} \times \mathbf{w})|$. This motivates the following:

Definition 2.3 We call the *scalar triple product* of three vectors $\mathbf{u}, \mathbf{v}$, and $\mathbf{w}$ the number given by

$$\mathbf{u} \cdot \mathbf{v} \times \mathbf{w}. \tag{2.2}$$

Note that we left out the brackets since $(\mathbf{u} \cdot \mathbf{v}) \times \mathbf{w}$ is meaningless.

Using (2.1), we can derive the expression for the scalar triple product when the vectors $\mathbf{u}, \mathbf{v}$, and $\mathbf{w}$ are given componentwise in terms of the orthonormal basis $(\mathbf{i}, \mathbf{j}, \mathbf{k})$ ordered in a right-handed sense. This yields:

$$\mathbf{u} \cdot \mathbf{v} \times \mathbf{w} = \det \begin{bmatrix} u_1 & u_2 & u_3 \\ v_1 & v_2 & v_3 \\ w_1 & w_2 & w_3 \end{bmatrix}.$$

This formula gives straightforwardly two nice properties of the scalar triple product:

$$\begin{aligned} \mathbf{u} \cdot \mathbf{v} \times \mathbf{w} &= \mathbf{u} \times \mathbf{v} \cdot \mathbf{w}, \\ \mathbf{u} \cdot \mathbf{v} \times \mathbf{w} &= \mathbf{v} \cdot \mathbf{w} \times \mathbf{u} = \mathbf{w} \cdot \mathbf{u} \times \mathbf{v}. \end{aligned}$$

Example 2.1 Show that if $\{a, b, c\}$ is an orthonormal set of vectors in $\mathbb{R}^3$, then

$$\mathbf{a} \cdot \mathbf{b} \times \mathbf{c} = \pm 1.$$

Solution: For $\sqrt{g} = 1$, it follows that

$$(\mathbf{a} \cdot \mathbf{b} \times \mathbf{c})^2 = \begin{vmatrix} a^1 & b^1 & c^1 \\ a^2 & b^2 & c^2 \\ a^3 & b^3 & c^3 \end{vmatrix}^2 = \begin{vmatrix} \mathbf{a} \cdot \mathbf{a} & \mathbf{b} \cdot \mathbf{b} & \mathbf{c} \cdot \mathbf{c} \\ \mathbf{b} \cdot \mathbf{a} & \mathbf{b} \cdot \mathbf{b} & \mathbf{b} \cdot \mathbf{c} \\ \mathbf{c} \cdot \mathbf{a} & \mathbf{c} \cdot \mathbf{b} & \mathbf{c} \cdot \mathbf{c} \end{vmatrix}.$$

Then

$$(\mathbf{a} \cdot \mathbf{b} \times \mathbf{c})^2 = \det \begin{bmatrix} 1 & 0 & 0 \\ 0 & 1 & 0 \\ 0 & 0 & 1 \end{bmatrix} = 1.$$

Let $(\bar{x}^1, \bar{x}^2, \bar{x}^3)$ be a rectangular Cartesian system of coordinates with orthonormal base vectors $(\mathbf{i}_1, \mathbf{i}_2, \mathbf{i}_3)$ and let (x^1, x^2, x^3) be a curvilinear system of coordinates with base vectors $(\mathbf{g}_1, \mathbf{g}_2, \mathbf{g}_3)$. For these coordinate systems, the vectors $\mathbf{g}_i$ and $\mathbf{i}_k$ are related as follows:

$$\mathbf{g}_i = J_i^k \, \mathbf{i}_k \quad \text{where} \quad J_i^k = \frac{\partial \bar{x}^k}{\partial x^i}. \tag{2.3}$$

Therefore,

$$g_{ij} = \mathbf{g}_i \cdot \mathbf{g}_j = \frac{\partial \bar{x}^k}{\partial x^i} \frac{\partial \bar{x}^r}{\partial x^j} \mathbf{i}_k \cdot \mathbf{i}_r = \sum_{k=1}^{3} \frac{\partial \bar{x}^k}{\partial x^i} \frac{\partial \bar{x}^k}{\partial x^j},$$

so that

$$\det[g_{ij}] = \det \left[\frac{\partial \bar{x}^k}{\partial x^i} \right] . \det \left[\frac{\partial \bar{x}^k}{\partial x^j} \right].$$

Denoting

$$g = \det[g_{ij}] = \mid g_{ij} \mid \quad \text{and} \quad \det \left[\frac{\partial \bar{x}^k}{\partial x^i} \right] = \left| \frac{\partial \bar{x}^k}{\partial x^i} \right|,$$

we can write

$$g = \mid g_{ij} \mid = \left| \frac{\partial \bar{x}^k}{\partial x^i} \right| \left| \frac{\partial \bar{x}^k}{\partial x^j} \right| \quad \text{or} \quad \sqrt{g} = \left| \frac{\partial \bar{x}^k}{\partial x^i} \right|. \tag{2.4}$$

Here, we assumed that the Jacobian of the transformation is positive.

At this point, we are ready to compute the volume dV of an infinitesimal parallelepiped with sides $\mathbf{g}_1 dx^1$, $\mathbf{g}_2 dx^2$, and $\mathbf{g}_3 dx^3$. Since

$$dV = \mathbf{g}_1 \cdot \mathbf{g}_2 \times \mathbf{g}_3 dx^1 dx^2 dx^3,$$

we can use (2.2), (2.3), and (2.4) to obtain

$$\mathbf{g}_1 \cdot \mathbf{g}_2 \times \mathbf{g}_3 = \left| \frac{\partial \bar{x}^k}{\partial x^i} \right|.$$

Thus,

$$dV = \sqrt{g}\, dx^1 dx^2 dx^3.$$

As an example, we have that cylindrical coordinates (x^i) and rectangular coordinates $(\bar{x}^k)$ are connected through

$$\begin{aligned} \bar{x}^1 &= x^1 \cos x^2, \\ \bar{x}^2 &= x^1 \sin x^2, \\ \bar{x}^3 &= x^3. \end{aligned}$$

Writing $x^1 = \rho$, $x^2 = \theta$, and $x^3 = z$, the formula for the volume element in this case is

$$dV = \rho\; d\rho\; d\theta\; dz.$$

Indeed, we have

$$\left| \frac{\partial \bar{x}^k}{\partial x^i} \right| = \det \begin{bmatrix} \cos x^2 & \sin x^2 & 0 \\ -x^1 \sin x^2 & x^1 \cos x^2 & 0 \\ 0 & 0 & 1 \end{bmatrix} = x^1 = \rho.$$

Example 2.2 The so-called paraboloidal coordinate system (x^i) is defined relatively to a rectangular Cartesian system $(\bar{x}^k)$ by means of the equations

$$\begin{aligned} \bar{x}^1 &= x^1 x^2 \cos x^3, \\ \bar{x}^2 &= x^1 x^2 \sin x^3, \\ \bar{x}^2 &= \frac{1}{2}\left[(x^1)^2 - (x^2)^2\right], \end{aligned}$$

where $x^1, x^2 \geq 0$, and $0 \leq x^3 \leq 2\pi$. Evaluate the corresponding volume element dV.

Solution: For this case,

$$\left|\frac{\partial \bar{x}^k}{\partial x^i}\right| = \det \begin{bmatrix} x^2 \cos x^3 & x^2 \sin x^3 & x^1 \\ x^1 \cos x^3 & x^1 \sin x^3 & -x^2 \\ -x^1 x^2 \sin x^3 & x^1 x^2 \cos x^3 & 0 \end{bmatrix}.$$

Evaluating the determinant yields

$$\left|\frac{\partial \bar{x}^k}{\partial x^i}\right| = x^1 x^2 \left((x^1)^2 + (x^2)^2\right)$$

so that

$$dV = x^1 x^2 \left((x^1)^2 + (x^2)^2\right) dx^1 dx^2 dx^3.$$

Now we consider the vector product and scalar triple product for arbitrary bases. For this purpose, let $\mathbf{A} = A^i \mathbf{g}_i$ and $\mathbf{B} = B^i \mathbf{g}_i$ be two vectors of the Euclidean vector space $\mathbf{E}$ expanded componentwise with respect to an arbitrary right-handed basis $(\mathbf{g}_1, \mathbf{g}_2, \mathbf{g}_3)$. Then, the vector product of $\mathbf{A}$ and $\mathbf{B}$ is

$$\mathbf{A} \times \mathbf{B} = \det \begin{bmatrix} \mathbf{g}_2 \times \mathbf{g}_3 & \mathbf{g}_3 \times \mathbf{g}_1 & \mathbf{g}_1 \times \mathbf{g}_2 \\ A^1 & A^2 & A^3 \\ B^1 & B^2 & B^3 \end{bmatrix}.$$

We can write this as

$$\mathbf{A} \times \mathbf{B} = e_{ijk} A^i B^j \mathbf{G}^k \tag{2.5}$$

where

$$\mathbf{G}^1 = \mathbf{g}_2 \times \mathbf{g}_3, \quad \mathbf{G}^2 = \mathbf{g}_3 \times \mathbf{g}_1, \quad \mathbf{G}^3 = \mathbf{g}_1 \times \mathbf{g}_2.$$

One can prove that the vectors $\{\mathbf{G}^1, \mathbf{G}^2, \mathbf{G}^3\}$ can be expressed in terms of the reciprocal basis $\{\mathbf{g}^1, \mathbf{g}^2, \mathbf{g}^3\}$ as follows:

$$\begin{aligned} \mathbf{G}^1 &= (\mathbf{g}_1 \times \mathbf{g}_2 \cdot \mathbf{g}_3)\mathbf{g}^1 = \sqrt{g}\mathbf{g}^1, \\ \mathbf{G}^2 &= (\mathbf{g}_1 \times \mathbf{g}_2 \cdot \mathbf{g}_3)\mathbf{g}^2 = \sqrt{g}\mathbf{g}^2, \\ \mathbf{G}^3 &= (\mathbf{g}_1 \times \mathbf{g}_2 \cdot \mathbf{g}_3)\mathbf{g}^3 = \sqrt{g}\mathbf{g}^3. \end{aligned}$$

Substituting this in (2.5), we have

$$\mathbf{A} \times \mathbf{B} = \epsilon_{ijk} A^i B^j \mathbf{g}^k,$$

where

$$\epsilon_{ijk} = \sqrt{g} e_{ijk}.$$

Concerning the scalar triple product, we obtain

$$\mathbf{A} \times \mathbf{B} \cdot \mathbf{C} = \epsilon_{ijk} A^i B^j C^k = \sqrt{g} \det \begin{bmatrix} A^1 & B^1 & C^3 \\ A^2 & B^2 & C^2 \\ A^3 & B^3 & C^3 \end{bmatrix}. \tag{2.6}$$

This formula yields easily that

$$\mathbf{A} \times \mathbf{B} \cdot \mathbf{C} = \mathbf{A} \cdot \mathbf{B} \times \mathbf{C}.$$

Also,

$$\mathbf{A} \cdot \mathbf{A} \times \mathbf{B} = \mathbf{B} \cdot \mathbf{A} \times \mathbf{B} = 0,$$

which means that the vector $\mathbf{A} \times \mathbf{B}$ is perpendicular to the plane determined by $\mathbf{A}$ and $\mathbf{B}$. Finally, taking $\mathbf{A} = \mathbf{g}_1 dx^1$, $\mathbf{B} = \mathbf{g}_2 dx^2$, and $\mathbf{C} = \mathbf{g}_3 dx^3$ in (2.6) and remembering that $dV = |\mathbf{A} \cdot \mathbf{B} \times \mathbf{C}|$, we obtain the expression for the volume element:

$$dV = \sqrt{g}\, dx^1 dx^2 dx^3.$$

2.2 Tensor algebra

The notion of tensor product of two or more vector spaces is a basic one in the theory of vector spaces and is essential to much of the following chapters. After starting off by considering the structure of the space $L(V, W)$ of all linear transformations between vector spaces and defining the notion of dual space V^* of a vector space V within this structure, we introduce the concept of the tensor product $\mathbf{v} \otimes \mathbf{u}$ of two vectors $\mathbf{v} \in V$ and $\mathbf{u} \in U$ as an element of $L(U^*, V)$. The concept of tensor itself is introduced by means of the idea of change of bases.

2.2.1 *Dual spaces*

Unless otherwise stated, we shall consider only real vector spaces, that is, vector spaces over the field $\mathbb{R}$ of real numbers. Also, all the propositions stated in this section are standard results whose proofs can be found in any linear algebra book (see, for instance, [Hoffman and Kunze, 1971]).

Definition 2.4 *Let V and W be vector spaces. A mapping $\mathbf{T}: V \to W$ is said to be a* linear transformation *if, for all $\mathbf{u}$, $\mathbf{v} \in V$ and $\alpha \in \mathbb{R}$,*

$$\mathbf{T}(\mathbf{u}+\mathbf{v}) = \mathbf{T}(\mathbf{u}) + \mathbf{T}(\mathbf{v}), \tag{2.7}$$

$$\mathbf{T}(\alpha\mathbf{u}) = \alpha\mathbf{T}(\mathbf{u}). \tag{2.8}$$

Sometimes, we will leave out the parenthesis and use the notation $\mathbf{Tu}$ instead of $\mathbf{T}(\mathbf{u})$. Examples of such transformations abound in mathematics. For instance, the so called projection from $\mathbb{R}^3$ to $\mathbb{R}^2$, defined by

$$\mathbf{P}(v^1, v^2, v^3) = (v^1, v^2), \quad \forall (v^1, v^2, v^3) \in \mathbb{R}^3,$$

and some other well known transformations on the plane, such as reflection with respect to a line and rotation about a point, are linear transformations.

Two basic properties of linear transformations are obtained by putting $\alpha = 0$ and $\alpha = -1$ in (2.8):

$$\mathbf{T}(\mathbf{0}) = \mathbf{0} \quad \text{and} \quad \mathbf{T}(-\mathbf{u}) = -\mathbf{T}(\mathbf{u}), \quad \forall \mathbf{u} \in V.$$

Observe that we are using the same symbol for the zero vector in the space V as well as W.

It can be easily shown that the set of all linear transformations from V to W, denoted $L(V, W)$, is itself a vector space, with addition and multiplication by a scalar defined by

$$(\mathbf{T}+\mathbf{S})(\mathbf{u}) = \mathbf{T}(\mathbf{u}) + \mathbf{S}(\mathbf{u}), \quad \forall \mathbf{T}, \mathbf{S} \in L(V, W) \text{ and } \mathbf{u} \in V,$$

$$(\alpha\mathbf{T})(\mathbf{u}) = \mathbf{T}(\alpha\mathbf{u}), \quad \forall \mathbf{T} \in L(V, W),\ \mathbf{u} \in V \text{ and } \alpha \in \mathbb{R}.$$

The *composition* (or *product*) of two transformations is the operation that associates to each pair $(\mathbf{T}, \mathbf{S}) \in L(U, W) \times L(V, U)$ an element $\mathbf{TS} \in L(V, W)$ such that $\mathbf{TS} = \mathbf{T} \circ \mathbf{S}$, that is, given any $\mathbf{S} \in L(V, U)$ and $\mathbf{T} \in L(U, W)$, we put

$$\mathbf{TS}(\mathbf{v}) = \mathbf{T}(\mathbf{S}(\mathbf{v})), \quad \forall \mathbf{v} \in V.$$

One can easily check that $\mathbf{TS}$ so defined is indeed linear. Also, it is immediate to show that the product satisfies the following properties:

P1 For all linear transformations

$$\mathbf{S} \in L(V, U),\ \mathbf{T} \in L(U, Y),\ \mathbf{U} \in L(Y, W)$$

on vector spaces, we have

$$\mathbf{U}(\mathbf{TS}) = (\mathbf{UT})\mathbf{S}.$$

P2 Consider $\mathbf{S}, \mathbf{T} \in L(V,U)$ and $\mathbf{U} \in L(U,W)$. Then,

$$\mathbf{U}(\mathbf{S}+\mathbf{T}) = \mathbf{US} + \mathbf{UT}.$$

P3 Consider $\mathbf{U} \in L(V,U)$ and $\mathbf{S}, \mathbf{T} \in L(U,W)$. Then,

$$(\mathbf{S}+\mathbf{T})\mathbf{U} = \mathbf{SU} + \mathbf{TU}.$$

In general, $\mathbf{ST} \neq \mathbf{TS}$, as shown by the case of $\mathbf{S} \in L(\mathbb{R}^3, \mathbb{R}^2)$ and $\mathbf{T} \in L(\mathbb{R}^2, \mathbb{R}^3)$ given by

$$\mathbf{S}(v^1, v^2, v^3) = (v^1, v^2) \quad \text{and} \quad \mathbf{T}(v^1, v^2) = (v^1, v^2, v^2)$$

for all $v^1, v^2, v^3 \in \mathbb{R}$. Indeed, notice that $\mathbf{ST} : \mathbb{R}^2 \to \mathbb{R}^2$ and $\mathbf{TS} : \mathbb{R}^3 \to \mathbb{R}^3$, which is enough to show that they are different mappings. Moreover,

$$\mathbf{ST}(v^1, v^2) = \mathbf{S}(v^1, v^2, v^2) = (v^1, v^2),$$
$$\mathbf{TS}(v^1, v^2, v^3) = \mathbf{T}(v^1, v^2) = (v^1, v^2, v^2).$$

We know that if a map $\mathbf{T} : V \to W$ is bijective (both injective and surjective), then it is invertible, i.e., there exists the inverse map $\mathbf{T}^{-1} : W \to V$, uniquely determined, such that $\mathbf{T}^{-1}\mathbf{T}$ is the identity map on V, denoted $\mathbf{I}_V$, and $\mathbf{T}\mathbf{T}^{-1}$ is the identity map on W, denoted $\mathbf{I}_W$. We have that if a linear transformation is invertible, then its inverse is also linear. More precisely, one can easily show the following result.

Proposition 2.1 . *(i) Let V and W be vector spaces. If $\mathbf{T} \in L(V,W)$ is bijective, then $\mathbf{T}^{-1} \in L(W,V)$.*

(ii) Let V, W, and U be vector spaces. If $\mathbf{T} \in L(U,W)$ and $\mathbf{S} \in L(V,U)$ are invertible, then $\mathbf{TS} \in L(V,W)$ is invertible and

$$(\mathbf{TS})^{-1} = \mathbf{S}^{-1}\mathbf{T}^{-1}.$$

We can write the components of a linear transformation in terms of a basis of $L(V,W)$. For this, let $\{\mathbf{a}_1, \mathbf{a}_2, \ldots, \mathbf{a}_n\}$ be a basis for V and $\{\mathbf{b}_1, \mathbf{b}_2, \ldots, \mathbf{b}_m\}$ be a basis for W. We define $n \times m$ linear transformations $\mathbf{E}^k_\alpha$ on $L(V,W)$ by putting

$$\mathbf{E}^k_\alpha \mathbf{a}_p := \delta^k_p \mathbf{b}_\alpha \tag{2.9}$$

for each $k, p = 1, 2, \ldots, n$ and $\alpha = 1, 2, \ldots, m$.

On the other hand, given any $\mathbf{T} \in L(V,W)$, we may expand $\mathbf{T}(\mathbf{a}_k) \in W$ as

$$\mathbf{T}(\mathbf{a}_k) = T^\alpha_k \mathbf{b}_\alpha. \tag{2.10}$$

Substituting (2.9) into (2.10), we obtain

$$\mathbf{T}(\mathbf{a}_k) = T_s^{\alpha}\mathbf{E}_{\alpha}^{s}\mathbf{a}_k.$$

Since $\{\mathbf{a}_1, \mathbf{a}_2, \ldots, \mathbf{a}_n\}$ is a basis for V, we can write

$$(\mathbf{T} - T_s^{\alpha}\mathbf{E}_{\alpha}^{s})(\mathbf{u}) = \mathbf{0}, \ \forall \mathbf{u} \in V,$$

which means that

$$\mathbf{T} = T_s^{\alpha}\mathbf{E}_{\alpha}^{s} \tag{2.11}$$

that is, the $n \times m$ linear transformations $\mathbf{E}_{\alpha}^{s}$ generate $L(V, W)$. It remains to show that they are linearly independent. For this, $T_s^{\alpha}\mathbf{E}_{\alpha}^{s} = \mathbf{0}$ implies that

$$\mathbf{0} = T_s^{\alpha}\mathbf{E}_{\alpha}^{s}(\mathbf{a}_k) = T_k^{\alpha}\mathbf{b}_{\alpha}$$

so that T_k^{α} must be all equal to zero, since $\mathbf{b}_1, \mathbf{b}_2, \ldots, \mathbf{b}_m$ are linearly independent.

Given the linear transformation $\mathbf{T}$, the real numbers T_s^{α}, with $s = 1, 2, \ldots, n$ and $\alpha = 1, 2, \ldots m$, that appear in (2.11) are called the *components of* $\mathbf{T}$ *in the basis* $\mathbf{E}_{\alpha}^{s}$ of $L(V, W)$. Note that, from (2.10), we have

$$\mathbf{T}(\mathbf{a}_k) = T_k^{\alpha}\mathbf{b}_{\alpha}.$$

We have just shown that the $n \times m$ linear transformations $\mathbf{E}_{\alpha}^{s}$ form a basis for $L(V, W)$. Hence, we have:

Proposition 2.2 *Suppose V and W are finite dimensional spaces. Then,*

$$\dim L(V, W) = \dim V \cdot \dim W. \tag{2.12}$$

The components that appear on (2.11) can be used to create a $n \times m$ matrix

$$[T_j^i]_{n\times m} = \begin{bmatrix} T_1^1 & T_1^2 & \cdots & T_1^m \\ T_2^1 & T_2^2 & \cdots & T_2^m \\ \vdots & \vdots & & \vdots \\ T_n^1 & T_n^2 & \cdots & T_n^m \end{bmatrix},$$

called the *matrix of the linear transformation* $\mathbf{T}$ *relative to the bases*$\{\mathbf{a}_1, \mathbf{a}_2, \ldots, \mathbf{a}_n\}$ *and* $\{\mathbf{b}_1, \mathbf{b}_2, \ldots, \mathbf{b}_m\}$. In the special case of the identity map $\mathbf{I}_V \in L(V, V)$ defined by $\mathbf{I}_V(\mathbf{v}) = \mathbf{v}$, its corresponding matrix relative to any basis for V is the identity matrix

$$I_{n\times n} = \begin{bmatrix} 1\,0\,0 & \cdots & 0 \\ 0\,1\,0 & \cdots & 0 \\ \vdots\,\vdots\,\vdots & \vdots & \vdots \\ 0\,0\cdots & 1 & 0 \\ 0\,0\cdots & 0 & 1 \end{bmatrix}.$$

Example 2.3 Let $\{\mathbf{a}_1, a_2, \ldots, a_n\}$ be a basis for V and take $v_i = \alpha_i^j \mathbf{a}_j$. Show that $\{\mathbf{v}_1, v_2, \ldots, v_n\}$ is linearly independent if and only if the matrix $[\alpha_i^j]_{n\times n}$ is invertible.

Solution. Suppose $\beta^i \in \mathbb{R}$, $i = 1, 2, \ldots n$, are such that $\beta^i \mathbf{v}_i = \mathbf{0}$. Then,

$$\begin{aligned} \mathbf{0} &= \beta^1 \left(\alpha_1^j \mathbf{a}_j\right) + \cdots + \beta^n \left(\alpha_n^j \mathbf{a}_j\right) \\ &= \left(\beta^i \alpha_i^1\right) \mathbf{a}_1 + \cdots + \left(\beta^i \alpha_i^n\right) \mathbf{a}_n. \end{aligned}$$

Since $\{\mathbf{a}_1, \mathbf{a}_2, \ldots, \mathbf{a}_n\}$ is linearly independent, we have $\beta^i \alpha_i^j = 0$, $j = 1, 2, \ldots, n$. This is a system of n linear equations in n unknowns $\beta^1, \beta^2, \ldots, \beta^n$ which, by Cramer's rule, has a nontrivial solution if and only if the matrix $[\alpha_i^j]_{n\times n}$ is nonsingular and the result follows.

The concept of linear transformations allows us to classify all finite dimensional vector spaces by telling that every abstract vector space V with $\dim V = n$ can be identified, in a precise sense, with the more concrete space $\mathbb{R}^n$. In order to do this characterization, first we need to introduce some definitions. Let V and W be vector spaces and consider $\mathbf{T} \in L(V, W)$. It can easily be shown that $\ker \mathbf{T} = \{\mathbf{v} \in V;\ \mathbf{T}\mathbf{v} = \mathbf{0}\}$, the *kernel* of $\mathbf{T}$, and $\operatorname{Im} \mathbf{T} = \{\mathbf{T}(\mathbf{v}); \mathbf{v} \in V\}$, the *image* of $\mathbf{T}$, are subspaces of V and W, respectively, and that $\dim \ker \mathbf{T} + \dim \operatorname{Im} \mathbf{T} = n$.

Definition 2.5
(i) A linear transformation $\mathbf{T} \in L(V, W)$ is called an*isomorphism* if $\mathbf{T}$ is bijective. In this case, we say that V is *isomorphic* to W.
(ii) $\mathbf{T} \in L(V, W)$ is called a *nonsingular transformation* if $\ker \mathbf{T} = \{\mathbf{0}\}$.

Note that if $\mathbf{T} \in L(V, W)$ is an isomorphism, then $\mathbf{T}^{-1} \in L(W, V)$ is also an isomorphism. So V is isomorphic to W if and only if W is isomorphic to V and we can say simply that V and W are isomorphic.

For linear transformations between finite dimensional vector spaces with the same dimension, the concepts of isomorphism and invertible and nonsingular transformations are all equivalent, so that to prove that a linear transformation is invertible all we need to do is to verify that it is either injective or surjective. This is a direct consequence of the following stronger result:

Proposition 2.3 *Let V and W be vector spaces with* $\dim V = \dim W = n$ *and let* $\mathbf{T} \in L(V, W)$. *Then, the following statements are equivalent:*

(1) $\mathbf{T}$ *is an isomorphism.*
(2) $\mathbf{T}$ *is invertible.*
(3) $\mathbf{T}$ *is injective.*
(4) $\mathbf{T}$ *is nonsingular.*
(5) rank $\mathbf{T} = \dim$ *and im*$\mathbf{T} = n$.
(6) $\mathbf{T}$ *is surjective.*
(7) If $\{\mathbf{v}_1, \mathbf{v}_2, \ldots, \mathbf{v}_n\}$ *is a basis for V, then* $\{\mathbf{T}(\mathbf{v}_1), \mathbf{T}(\mathbf{v}_2), \ldots, \mathbf{T}(\mathbf{v}_n)\}$ *is a basis for W.*

Now we are in condition to have the characterization of all finite dimensional vector spaces. Indeed, let $\{\mathbf{v}_1, \mathbf{v}_2, \ldots, \mathbf{v}_n\}$ be a basis for a n-dimensional (real) vector space V. Define a transformation $\mathbf{T} : V \to \mathbb{R}^n$ by putting

$$\mathbf{T}\mathbf{v} := (\alpha^1, \alpha^2, \ldots, \alpha^n) \tag{2.13}$$

for each $\mathbf{v} \in V$, $\mathbf{v} = \alpha^i \mathbf{v}_i$, where $\alpha^i \in \mathbb{R}$, $i = 1, 2, \ldots, n$, are uniquely determined. Then, we have:

Proposition 2.4 *If V is any vector space with* $\dim V = n$, *then V is isomorphic to* $\mathbb{R}^n$.

Example 2.4 Consider $\alpha \in R$ and $T_\alpha \in L(R^2, R^2)$ defined by

$$\mathbf{T}_\alpha(v^1, v^2) = (v^1 \cos\alpha - v^2 \sin\alpha, v^1 \sin\alpha + v^2 \cos\alpha).$$

Show that T_α is an isomorphism.

Solution. It is more convenient to express $\mathbf{T}_\alpha$ in the following manner:

$$\mathbf{T}_\alpha\left(\begin{Bmatrix} v^1 \\ v^2 \end{Bmatrix}\right) = \begin{Bmatrix} v^1 \cos\alpha - v^2 \sin\alpha \\ v^1 \sin\alpha + v^2 \cos\alpha \end{Bmatrix} = A_\alpha \begin{Bmatrix} v^1 \\ v^2 \end{Bmatrix},$$

where A_α is the 2×2 matrix

$$A_\alpha = \begin{bmatrix} \cos\alpha & -\sin\alpha \\ \sin\alpha & \cos\alpha \end{bmatrix}.$$

To show that $\mathbf{T}_\alpha$ is an isomorphism is equivalent to show that the matrix A_α is invertible. But $\det A_\alpha = \cos^2\alpha + \sin^2\alpha = 1 \neq 0$, so that A_α is invertible and $A_\alpha A_\alpha^{-1} = I_{2\times 2}$ gives

$$A_\alpha^{-1} = \begin{bmatrix} \cos\alpha & \sin\alpha \\ -\sin\alpha & \cos\alpha \end{bmatrix}.$$

Let V be a vector space over the the field $\mathbb{R}$. The vector space $L(V,\mathbb{R})$ of all linear transformations from V into the vector space $\mathbb{R}$ turns up to be of great importance in the most diverse applications of mathematics because sometimes we can profit from converting relations in the original vector space V into statements applying to $L(V,\mathbb{R})$. We call such transformations *linear functionals* and we refer to this process as dualization.

Definition 2.6 Let V be a vector space (over the field $\mathbb{R}$). We say that f is a *linear functional* on V if $f \in L(V,\mathbb{R})$. In this case, we say that $L(V,\mathbb{R})$ is the *dual space* of V.

Usually, we denote $L(V,\mathbb{R})$ by V^* and their elements by $\mathbf{u}^*, \mathbf{v}^*, \mathbf{w}^*, \ldots$. Also, we may call *contravariant vectors* the elements of V to distinguish them from the elements of V^*, which we may call *covariant vectors.* Or else, we can call vectors the elements of V and covectors the elements of V^*.

It is easy to check that the following maps are examples of linear functionals:

(1) $f : \mathbb{R}^n \to \mathbb{R}$ defined by

$$f(x_1, \ldots, x_n) = c_1 x_1 + \cdots + c_n x_n$$

where $c_1, \ldots, c_n \in \mathbb{R}$.

(2) $f : C[0,1] \to \mathbb{R}$ defined by

$$f(x) = \int_0^1 x(t)d\alpha(t),$$

where we are using $C[0,1]$ to denote the vector space of all continuous functions from the closed interval $[0,1]$ into $\mathbb{R}$ and $\alpha(\cdot)$ is a function of bounded variation defined on $[0,1]$.

(3) Let $\mathbf{y} \in \mathbb{R}^n$. Define $f_y : \mathbb{R}^n \to \mathbb{R}$ by

$$f_y(\mathbf{x}) = \mathbf{x}\cdot\mathbf{y}, \ \ \forall\, \mathbf{x} \in \mathbb{R}^n,$$

where $\cdot$ denotes the dot product in $\mathbb{R}^n$.

When V is finite-dimensional ($\dim V = n$), we have from (2.12) that $\dim V^* = \dim L(V, \mathbb{R}) = n \cdot 1 = n$, from which it follows the proposition:

Proposition 2.5 *The spaces V and V^* are isomorphic.*

Let V be a n-dimensional vector space and $\{\mathbf{e}_1, \ldots, \mathbf{e}_n\}$ a basis for V. Then, to any given set of scalars $\{w_1, \ldots, w_n\}$, there corresponds a mapping $\mathbf{w}^* : V \to \mathbb{R}$, defined by

$$\mathbf{w}^*(\mathbf{e}_i) = w_i. \tag{2.14}$$

Note that in order to define a linear transformation on V it suffices to define it for the basis for V. Thus, we have that $\mathbf{w}^* \in V^*$ and that

$$\mathbf{w}^*(\mathbf{v}) = v^i w_i = v^1 w_1 + \cdots v^n w_n,$$

for each $\mathbf{v} = v^1\mathbf{e}_1 + \cdots + v^n\mathbf{e}_n \in V$.

Now, we consider $\{\mathbf{e}^{*1}, \ldots, \mathbf{e}^{*n}\} \subset V^*$ by putting

$$\mathbf{e}^{*i}(\mathbf{e}_j) = \delta^i_j \tag{2.15}$$

for all $i, j = 1, \ldots, n$. These conditions determine each $\mathbf{e}^{*i}$ uniquely. Moreover, we have the following result.

Proposition 2.6 *Let $\{\mathbf{e}_1, \ldots, \mathbf{e}_n\}$ be a basis for a n -dimensional vector space V. Then, $\{\mathbf{e}^{*1}, \ldots, \mathbf{e}^{*n}\}$ defined by (2.15) is a basis for V^*.*

■ **Proof:** Since $\dim V^* = n$, it suffices to prove that $\{\mathbf{e}^{*1}, \ldots, \mathbf{e}^{*n}\}$ is linearly independent. Thus, let $\alpha_1, \ldots, \alpha_n$ be real numbers such that $\alpha_i \mathbf{e}^{*i} = \mathbf{0}$. Then, for $j = 1, \ldots, n$, we have $\alpha_i \mathbf{e}^{*i}(\mathbf{e}_j) = 0$, so that $\alpha_i \delta^i_j = 0$. From this, it follows that $\alpha_i = 0$, for $i = 1, \ldots, n$. □

Definition 2.7 Let $\{\mathbf{e}_1, \ldots, \mathbf{e}_n\}$ be a basis of a n-dimensional vector space V. We define the *dual basis* of $\{\mathbf{e}_1, \ldots, \mathbf{e}_n\}$ to be the basis $\{\mathbf{e}^{*1}, \ldots, \mathbf{e}^{*n}\}$ of V^* such that the conditions (2.15) hold.

We can have another useful representation for linear functionals. Let us start with the concept of bilinear functions.

Definition 2.8 Let V, U, and W be vector spaces. A map $f : V \times U \to W$ is *bilinear* if

$$f(\mathbf{v}_1 + \mathbf{v}_2, \mathbf{w}) = f(\mathbf{v}_1, \mathbf{w}) + f(\mathbf{v}_2, \mathbf{w})$$
$$f(\mathbf{v}, \mathbf{w}_1 + \mathbf{w}_2) = f(\mathbf{v}, \mathbf{w}_1) + f(\mathbf{v}, \mathbf{w}_2)$$

and

$$f(\alpha\mathbf{v}, \mathbf{w}) = f(\mathbf{v}, \alpha\mathbf{w}) = \alpha f(\mathbf{v}, \mathbf{w}),$$

for all scalars α and all vectors $\mathbf{v}, \mathbf{v}_1, \mathbf{v}_2, \mathbf{w}, \mathbf{w}_1, \mathbf{w}_2$ in V. When $W = \mathbb{R}$, we say that f is a *bilinear functional.*

The set of all bilinear maps from $V \times U$ into W is denoted as $L(V \times U, W)$ and has the structure of a real vector space, with the usual operations of addition and multiplication by a scalar.

Thus, a map $f = f(\mathbf{v}, \mathbf{w})$ is bilinear if it is linear in $\mathbf{v}$ when $\mathbf{w}$ is held fixed, and is linear in $\mathbf{w}$ when $\mathbf{v}$ is held fixed. The vector product in ordinary three-dimensional space is a familiar example of a bilinear map when $V = U = W$. An example of a bilinear functional that we have already come across is the inner product.

A bilinear functional $\mathbf{f}$ defined on $V \times V$ is called *symmetric* if $\mathbf{f}(u, v) = \mathbf{f}(v, u)$ for all u, v in V, and *anti-symmetric* (or skew-symmetric) if $\mathbf{f}(u, v) = -\mathbf{f}(v, u)$ for all u, v in V. We have that a bilinear functional $\mathbf{f} : V \times V \to \mathbb{R}$ with the property that $\mathbf{f}(u, u) = 0$ for all u in V is necessarily anti-symmetric.

Consider two n-dimensional vector spaces V and U and take $\mathbf{T} \in L(V, U)$. Let $\{\mathbf{e}_1, \ldots, \mathbf{e}_n\}$ and $\{\mathbf{f}_1, \ldots, \mathbf{f}_n\}$ be bases for V and U, respectively. Since $T(\mathbf{e}_k) \in U$, we can write

$$T(\mathbf{e}_k) = T^i_k \mathbf{f}_i. \tag{2.16}$$

The $m \times n$ scalars T^i_k are called *the components of linear transformation* T for the bases $\{\mathbf{e}_1, \ldots, \mathbf{e}_n\}$ and $\{\mathbf{f}_1, \ldots, \mathbf{f}_n\}$. When it turns out necessary to distinguish such components T^i_k from other components of $\mathbf{T}$ (tensor components), we shall indicate explicitly the bases by using brackets, that is,

$$T^i_k = [T^i_k]^{\mathbf{f}}_{\mathbf{e}}.$$

Now, let $\{\mathbf{f}^{*1}, \ldots, \mathbf{f}^{*n}\}$ be the dual basis for $\{\mathbf{f}_1, \ldots, \mathbf{f}_n\}$. Applying $\mathbf{f}^{*j}$ to both sides of Eq. (2.16), we have

$$\mathbf{f}^{*j}(T(\mathbf{e}_k)) = T^i_k \mathbf{f}^{*j}(\mathbf{f}_i) = T^i_k \delta^j_i = T^j_k,$$

due to the definition of dual basis. Therefore, $T^j_k = \mathbf{f}^{*j}(T(\mathbf{e}_k))$ are the components of $\mathbf{T}$ with respect to the bases $\{\mathbf{e}_1, \ldots, \mathbf{e}_n\}$ and $\{\mathbf{f}^{*1}, \ldots, \mathbf{f}^{*n}\}$.

2.2.2 *Duality and inner product*

The linearity property inherently enjoyed by each element $\mathbf{v}^* \in V^*$ allows us to view $\mathbf{v}^*(\mathbf{u})$, for any $\mathbf{u} \in V$, as the result of a bilinear functional acting on the pair $(\mathbf{v}^*, \mathbf{u}) \in V^* \times V$. Since bilinear functionals have a product structure coming out from the properties (B1)–(B4) below, it is convenient sometimes to denote a bilinear functional by the bracket notation $< ., . >$ in order to bring out this product structure. Therefore, we shall identify the linear operation on V^* with the bilinear map

$$\begin{aligned} <.,.>: V^* \times V &\to \mathbb{R} \\ (\mathbf{v}^*, \mathbf{u}) &\longmapsto <\mathbf{v}^*, \mathbf{u}>:= \mathbf{v}^*(\mathbf{u}) \end{aligned} \tag{2.17}$$

connecting the elements of the spaces V^* and V and enjoying the following properties:

(B1) $<\alpha\mathbf{v}^* + \beta\mathbf{u}^*, \mathbf{w}>= \alpha <\mathbf{v}^*, \mathbf{w}> +\beta <\mathbf{u}^*, \mathbf{w}>$, for all scalars α, β, and $\mathbf{v}^*, \mathbf{u}^* \in V^*$ and $\mathbf{w} \in V$.

(B2) $<\mathbf{v}^*, \alpha\mathbf{u} + \beta\mathbf{v}>= \alpha <\mathbf{v}^*, \mathbf{u}> +\beta <\mathbf{v}^*, \mathbf{v}>$, for all scalar α, β, and $\mathbf{v}^* \in V^*$ and $\mathbf{u}, \mathbf{v} \in V$.

(B3) For any given $\mathbf{v}^* \in V^*$,

$$<\mathbf{v}^*, \mathbf{w}>= 0, \ \ \forall \mathbf{w} \in V, \ \text{ if and only if } \ \mathbf{v}^* = \mathbf{0}.$$

(B4) For any given $\mathbf{v} \in V$,

$$<\mathbf{v}^*, \mathbf{v}>= 0, \ \ \forall \mathbf{v}^* \in V^*, \ \text{ if and only if } \ \mathbf{v} = \mathbf{0}.$$

Because of this identification, we say that the two spaces V^* and V are in *duality*. We should note that it is important to distinguish between the bracket operation $<.,.>$ on $V^* \times V$ and the inner product on V, which is a scalar valued bilinear mapping defined on $V \times V$.

Let V be a vector space with an inner product $\cdot$ and $\mathbf{v} \in V$ be fixed. It is an easy task to prove that the functional $f_{\mathbf{v}} : V \to \mathbb{R}$, defined by

$$f_{\mathbf{v}}(\mathbf{u}) = \mathbf{v} \cdot \mathbf{u}, \quad \forall \mathbf{u} \in V$$

is linear, so that $f_{\mathbf{v}} \in V^*$. More conveniently, we can say that $f_{\mathbf{v}} = \mathbf{v}^*$, for some $\mathbf{v}^* \in V^*$. Taking into account (2.17), we have

$$f_{\mathbf{v}}(\mathbf{u}) = \mathbf{v}^*(\mathbf{u}) = <\mathbf{v}^*, \mathbf{u}>$$

so that, for this $\mathbf{v}^*$,

$$\mathbf{v} \cdot \mathbf{u} = <\mathbf{v}^*, \mathbf{u}>. \tag{2.18}$$

Therefore, to any given $\mathbf{v} \in V$ there corresponds an element $\mathbf{v}^* \in V^*$ such that (2.18) is satisfied, for all $\mathbf{u} \in V$. Moreover, we know from proposition 8 that V and V^* are isomorphic whenever V is finite-dimensional.

Proposition 2.7 *Let V be a Euclidean vector space and consider the mapping $G : V \to V^*$ defined by means of the inner product through the formula*

$$<G(\mathbf{v}), \mathbf{u}>= \mathbf{v} \cdot \mathbf{u}, \quad \forall \mathbf{v}, \mathbf{u} \in V. \tag{2.19}$$

Then, G is an isomorphism between V and V^*.

The following straightforward proposition relates the reciprocal and dual bases by showing that they can be identified.

Proposition 2.8 *Suppose $\{\mathbf{e}^1, \ldots, \mathbf{e}^n\}$ and $\{\mathbf{e}^{*1}, \ldots, \mathbf{e}^{*n}\}$ are, respectively, the reciprocal and the dual bases for an arbitrary basis $\{\mathbf{e}_1, \ldots, \mathbf{e}_n\}$ in an Euclidean vector space V. Then,*

$$G\mathbf{e}^i = \mathbf{e}^{*i}, \quad i = 1, \ldots, n, \tag{2.20}$$

where G is the isomorphism defined by (2.19). Moreover, for each given $\mathbf{v}^* = v_i^* \mathbf{e}^{*i} \in V^*$, there exists a $\mathbf{v} = v_i \mathbf{e}^{\mathbf{i}} \in V$ such that $v_i^* = v_i$.

Recall that (2.18) implies that $< ., . >$ is indeed an inner product. Thus, we can write

$$\begin{aligned}
&< \mathbf{v}^*, \mathbf{e}_i > = < v_i^* \mathbf{e}^{*i}, \mathbf{e}_i > = v_i^* < \mathbf{e}^{*i}, \mathbf{e}_i > = v_i^* \\
&< \mathbf{e}^{*i}, \mathbf{v} > = < \mathbf{e}^{*i}, v^i \mathbf{e}_i > = v^i < \mathbf{e}^{*i}, \mathbf{e}_i > = v^i \\
&< \mathbf{v}^*, \mathbf{v} > = < v_i^* \mathbf{e}^{*i}, v^i \mathbf{e}_i > = v_i^* v^i < \mathbf{e}^{*i}, \mathbf{e}_i > = v_i^* v^i
\end{aligned}$$

and we have the following generalization from inner product spaces to vector spaces in general:

$$< \mathbf{v}^*, \mathbf{e}_i > = v_i^*, \quad < \mathbf{e}^{*i}, \mathbf{v} > = v^i, \quad < \mathbf{v}^*, \mathbf{v} > = v_i^* v^i$$

for $\mathbf{v}^* = v_i^* \mathbf{e}^{*i}$ and $\mathbf{v} = v^i \mathbf{e}_i$.

The following result shows that the second dual space $V^{**} = L(L(V, \mathbb{R}), \mathbb{R})$ can be identified with V.

Proposition 2.9 *Let V be an Euclidean vector space. Then, V and V^{**} are isomorphic.*

Any linear transformation $\mathbf{T} \in L(V, U)$ induces, in a unique manner, another transformation $\mathbf{T}^T \in L(U^*, V^*)$, which we call *the transpose of* $\mathbf{T}$, by means of the following procedure. Define $\mathbf{T}^T$ from U^* to V^* by setting, for all $\mathbf{u}^* \in U^*$, the composition $\mathbf{T}^T(\mathbf{u}^*) := \mathbf{u}^* \circ \mathbf{T}$ or, leaving out the parentheses and the composition notation,

$$\mathbf{T}^T \mathbf{u}^* := \mathbf{u}^* \mathbf{T},$$

that is,

$$\mathbf{T}^T \mathbf{u}^* (\mathbf{v}) = \mathbf{u}^* (\mathbf{T}\mathbf{v}), \quad \forall \mathbf{v} \in V. \tag{2.21}$$

Note that $\mathbf{T}^T\mathbf{u}^* \in V^*$, because it is the composition of two linear transformations. Moreover, $\mathbf{T}^T$ is a linear transformation from U^* into V^*, since

$$\begin{aligned}\mathbf{T}^T(\alpha\mathbf{u}_1^* + \mathbf{u}_2^*) &= (\alpha\mathbf{u}_1^* + \mathbf{u}_2^*) \circ \mathbf{T}\\ &= \alpha\mathbf{u}_1^* \circ \mathbf{T} + \mathbf{u}_2^* \circ \mathbf{T}\\ &= \alpha\mathbf{T}^T\mathbf{u}_1^* + \mathbf{T}^T\mathbf{u}_2^*\end{aligned}$$

for all $\mathbf{u}_1^*, \mathbf{u}_2^* \in U^*$ and $\alpha \in \mathbb{R}$.

As for the uniqueness, if $\mathbf{T}_1^T, \mathbf{T}_2^T \in L(U^*, V^*)$ are transposes of $\mathbf{T}$, then

$$\mathbf{u}^*(\mathbf{T}\mathbf{v}) = \mathbf{T}_1^T\mathbf{u}^*(\mathbf{v}) = \mathbf{T}_2^T\mathbf{u}^*(\mathbf{v})$$

for all $\mathbf{v} \in V$ and $\mathbf{u}^* \in U^*$. Therefore, $(\mathbf{T}_1^T\mathbf{u}^* - \mathbf{T}_2^T\mathbf{u}^*)(\mathbf{v}) = \mathbf{0}$, which gives that $(\mathbf{T}_1^T - \mathbf{T}_2^T)\mathbf{u}^* = \mathbf{0}$ (from the arbitrariness of $\mathbf{v}$) and that $\mathbf{T}_1^T = \mathbf{T}_2^T$ (from the arbitrariness of $\mathbf{u}^*$).

In view of the bracket notation introduced by (2.17), formula (2.21) can be rewritten as

$$< \mathbf{u}^*, \mathbf{T}\mathbf{v} >=< \mathbf{T}^T\mathbf{u}^*, \mathbf{v} > \quad , \quad \forall \mathbf{v} \in V \quad , \quad \forall \mathbf{u}^* \in U^*.$$

So, for inner product spaces, we define:

Definition 2.9 Let V and U be vector spaces with inner product $< .,. >_V$ and $< .,. >_U$, respectively.

Then, given a linear transformation $\mathbf{T} \in L(V, U)$, we define the *transpose* of $\mathbf{T}$ and denote $\mathbf{T}^T$ to be a linear transformation on $L(U, V)$ such that

$$< \mathbf{T}\mathbf{v}, \mathbf{u} >_U=< \mathbf{v}, \mathbf{T}^T\mathbf{u} >_V, \quad \forall\, \mathbf{v} \in V, \ \forall\, \mathbf{u} \in U.$$

The transpose enjoys the following properties:

Proposition 2.10 .

(1) For all $\mathbf{T}, \mathbf{S} \in L(V, U)$ and $\alpha \in \mathbb{R}$, we have that $(\mathbf{T}+\mathbf{S})^T, (\alpha\mathbf{T})^T \in L(U, V)$ are such that

$$(\mathbf{T} + \mathbf{S})^T = \mathbf{T}^T + \mathbf{S}^T \qquad \textit{and} \qquad (\alpha\mathbf{T})^T = \alpha\mathbf{T}^T.$$

(2) For all $\mathbf{T} \in L(V, U)$ and $\mathbf{S} \in L(U, W)$, we have that $(\mathbf{S}\mathbf{T})^T \in L(W, V)$ with the property

$$(\mathbf{S}\mathbf{T})^T = \mathbf{T}^T\mathbf{S}^T.$$

(3) For all $\mathbf{T} \in L(V, U)$,

$$(\mathbf{T}^T)^T = \mathbf{T}.$$

We say that $\mathbf{T} \in L(V,V)$ is *symmetric* if $\mathbf{T} = \mathbf{T}^T$, and we say that $\mathbf{T} \in L(V,V)$ is *skew-symmetric* if $\mathbf{T} = -\mathbf{T}^T$. Note that, given any linear transformation $\mathbf{T} \in L(V,V)$, it can always be expressed as a sum of a symmetric linear transformation and a skew-symmetric one, that is, $\mathbf{T} = \mathbf{S} + \mathbf{A}$, where

$$\mathbf{S} = \frac{1}{2}\left(\mathbf{T} + \mathbf{T}^T\right) \quad \text{and} \quad \mathbf{A} = \frac{1}{2}\left(\mathbf{T} - \mathbf{T}^T\right)$$

with $\mathbf{S}$ being symmetric and $\mathbf{A}$ skew-symmetric.

Example 2.5 Consider an inner product space V. Show that $T \in L(V,V)$ is skew-symmetric if and only if $\mathbf{v} \cdot T\mathbf{v} = 0,\ \forall\ \mathbf{v} \in V$.

Solution: Suppose $\mathbf{T}$ is skew-symmetric. Then, for any $\mathbf{v} \in V$, we have that

$$\mathbf{v} \cdot \mathbf{T}\mathbf{v} = \mathbf{T}^T\mathbf{v} \cdot \mathbf{v} = (-\mathbf{T}\mathbf{v}) \cdot \mathbf{v} = -(\mathbf{v} \cdot \mathbf{T}\mathbf{v}),$$

from which it follows that $\mathbf{v} \cdot \mathbf{T}\mathbf{v} = 0$.
Conversely, if $\mathbf{v} \cdot \mathbf{T}\mathbf{v} = 0,\ \forall \mathbf{v} \in V$, then for all $\mathbf{u}, \mathbf{v} \in V$ we have

$$\begin{aligned} (\mathbf{u} - \mathbf{v}) \cdot \mathbf{T}(\mathbf{u} - \mathbf{v}) &= 0, \\ \mathbf{u} \cdot \mathbf{T}\mathbf{u} - \mathbf{u} \cdot \mathbf{T}\mathbf{v} - \mathbf{v} \cdot \mathbf{T}\mathbf{u} + \mathbf{v} \cdot \mathbf{T}\mathbf{v} &= 0, \\ -\mathbf{u} \cdot \mathbf{T}\mathbf{v} - \mathbf{v} \cdot \mathbf{T}\mathbf{u} &= 0. \end{aligned}$$

Thus, $\mathbf{u} \cdot \mathbf{T}\mathbf{v} = -\mathbf{v} \cdot \mathbf{T}\mathbf{u}$.

2.2.3 *Tensors*

In physics, some quantities, such as the state of stress of a body at a given point, can be represented by what we call tensors. Tensors are abstract entities having a set of components that are functions of position in n-dimensional Euclidean space. So, if a point has n coordinates $(x^1, x^2, \ldots, x^n)$ in some coordinate system and corresponding coordinates $(\overline{x}^1, \overline{x}^2, \ldots, \overline{x}^n)$ in another coordinate system, then a set of n components A^i; $i = 1, 2, \ldots, n$, that are functions of the n coordinates x^i will become a set of n components $\overline{A}^i$ that are functions of the coordinates $\overline{x}^i$ after a change of coordinates from the first to the second system.

A *contravariant tensor of order 1* is a set of n components A^i satisfying, for each i, the following transformation law:

$$\overline{A}^i = \sum_{r=1}^{n} \frac{\partial \overline{x}^i}{\partial x^r} A^r \triangleq \frac{\partial \overline{x}^i}{\partial x^r} A^r, \text{ that is, } N_r^i = \frac{\partial \overline{x}^i}{\partial x^r} \text{ and } \overline{A}^i = N_r^i A^r.$$

A *covariant tensor of order 1* is a set of n components A_i satisfying, for each i, the relation

$$\overline{A}_i = \frac{\partial x^r}{\partial \overline{x}^i} A_r \Longleftrightarrow \overline{A}_i = J_i^r A_r.$$

A *contravariant tensor of order 2* is a set of n^2 components A^{ij} satisfying, for all i and j, the relation

$$\overline{A}^{ij} = \sum_{r=1}^{n} \sum_{s=1}^{n} \frac{\partial \overline{x}^i}{\partial x^r} \frac{\partial \overline{x}^j}{\partial x^s} A^{rs} \quad \Longleftrightarrow \quad T^{ij} = T^{kj} N_k^{\ i} N_j^{\ j}.$$

A *covariant tensor of order 2* is a set of n^2 components A_{ij} satisfying, for all i and j, the relation

$$\overline{A}_{ij} = \sum_{r=1}^{n} \sum_{s=1}^{n} \frac{\partial x^r}{\partial \overline{x}^i} \frac{\partial x^s}{\partial \overline{x}^j} A_{rs} \quad \Longleftrightarrow \quad T_{ij} = T_{rs} J^r_{\ i} J^s_{\ j}.$$

Tensors of higher order are similarly defined. Note that tensors of order 1 are vectors. Thus, we define a *tensor of order zero* to be a scalar.

A *tensor of order two* (or a second-order tensor) is a linear function that maps any vector into a vector or a bilinear functional. Thus, any product (or *diad*) $\mathbf{a} \otimes \mathbf{b}$ is a tensor of order two if it is understood that $\mathbf{a} \otimes \mathbf{b}$ is a linear mapping into vectors. A *tensor of order one* is a vector or a linear functional that maps any vector into a scalar. A *tensor of order zero* is a scalar. We note the following facts.

(1) A linear combination of dyads with scalar coefficients

$$\mathbf{D} = \alpha \mathbf{a} \otimes \mathbf{b} + \beta \mathbf{c} \otimes \mathbf{d} + \gamma \mathbf{e} \otimes \mathbf{f} + \kappa \mathbf{g} \otimes \mathbf{h}$$

is a tensor of order 2, called a *dyadic* (see [Reddy and Rasmussen, 1982; Reddy, 2013])

(2) For any vector $\mathbf{v}$ in a given vector space V, the linear mapping

$$\begin{aligned} \mathbf{L}_v : V &\longrightarrow V \\ \mathbf{u} &\longmapsto \mathbf{L}_v \mathbf{u} = \mathbf{v} \times \mathbf{u}, \end{aligned}$$

where $\times$ denotes the cross product, is also a tensor of order 2.

(3) Let $\{\mathbf{e}_1, \mathbf{e}_2, \ldots, \mathbf{e}_n\}$ and $\{\mathbf{f}_1, \mathbf{f}_2, \ldots, \mathbf{f}_m\}$ be the given bases of V and W, respectively. Then, every second-order tensor $\mathbf{T} \in V \otimes W$ can be written in the form

$$\mathbf{T} = \mathbf{e}_k \otimes \mathbf{w}^k,$$

where $\mathbf{w}^k \in W$ is such that $\mathbf{w}^k = T^{ks}\mathbf{f}_s$, for $k = 1, 2, \ldots, n$ and $s = 1, 2, \ldots, m$.

We formalize the concept of tensor in the following definition.

Definition 2.10 Let p and q be nonnegative integers and V be a n-dimensional vector space.

(1) A *tensor* $\mathbf{T}$ *of order* (p,q) *on* V is a multilinear functional

$$\mathbf{T} : \underbrace{V^* \times V^* \times \cdots \times V^*}_{p \text{ times}} \times \underbrace{V \times V \times \cdots \times V}_{q \text{ times}} \longrightarrow \mathbb{R}.$$

We denote the space of all tensors of order (p,q) on V by $T^p_q(V)$.

(2) A *contravariant tensor of order* p, or simply *a tensor of order* $(p, 0)$, is a p-linear functional

$$\mathbf{T} : \underbrace{V^* \times V^* \times \cdots \times V^*}_{p \text{ times}} \longrightarrow \mathbb{R}.$$

We denote by $T^p(V)$ the space of all such functionals.

(3) A *covariant tensor of order* q, or simply *a tensor of order* $(0, q)$, is a q-linear functional

$$\mathbf{T} : \underbrace{V \times V \times \cdots \times V}_{q \text{ times}} \longrightarrow \mathbb{R}.$$

We denote by $T_q(V)$ the space of all such functionals.

By convention, we use the following notation:

$$\begin{aligned} T^0_0(V) &= \mathbb{R}, \\ T^1(V) &= L(V^*, \mathbb{R}) = V^{**} \sim V, \\ T_1(V) &= L(V, \mathbb{R}) = V^*. \end{aligned}$$

Example 2.6 Let V be a finite dimensional vector space. Show that, for $v \in V$, $v^* \in V^*$ and any real number $\lambda \neq 0$, we have

$$\mathbf{v} \otimes \mathbf{v}^* = (\lambda \mathbf{v}) \otimes \left(\frac{1}{\lambda}\mathbf{v}^*\right).$$

Solution: We have

$$\begin{aligned}(\lambda \mathbf{v}) \otimes \left(\frac{1}{\lambda}\mathbf{v}^*\right)(\mathbf{u}^*, \mathbf{u}) &= <\mathbf{u}^*, \lambda\mathbf{v}> . <\frac{1}{\lambda}\mathbf{v}^*, \mathbf{u}> \\ &= \lambda\frac{1}{\lambda} <\mathbf{u}^*, \mathbf{v}> . <\mathbf{v}^*, \mathbf{u}> \\ &= <\mathbf{u}^*, \mathbf{v}> . <\mathbf{v}^*, \mathbf{u}> \\ &= (\mathbf{v} \otimes \mathbf{v}^*)(\mathbf{u}^*, \mathbf{u}).\end{aligned}$$

2.2.4 *Tensor product*

The notion of a tensor product of two or more vector spaces is a basic one in the theory of vector spaces and is essential to much of the following chapters. But, before introducing this concept, we start off with the concept of tensor product of vectors. Notice that $\mathbf{u}^*(\mathbf{u}) \in \mathbb{R}$, whenever $\mathbf{u}^* \in U^* = L(U, \mathbb{R})$ and $\mathbf{u} \in U$.

Definition 2.11 Let V and U be two finite-dimensional vector spaces. For any pair of vectors $(\mathbf{v}, \mathbf{u}) \in V \times U$, we call the *tensor product of* $\mathbf{v}$ *and* $\mathbf{u}$, and denote $\mathbf{v} \otimes \mathbf{u}$, the mapping $\mathbf{v} \otimes \mathbf{u} \in L(U^*, V)$ defined by

$$\begin{aligned}\mathbf{v} \otimes \mathbf{u} \ &: U^* \longrightarrow V \\ &\mathbf{u}^* \longmapsto \mathbf{v} \otimes \mathbf{u}\ (\mathbf{u}^*) := \mathbf{u}^*(\mathbf{u})\ (\mathbf{v}).\end{aligned}$$

We shall use the notation $\mathbf{v} \otimes \mathbf{u}\ (\mathbf{u}^*) = \mathbf{u}^*(\mathbf{u})\ (\mathbf{v}) = <\mathbf{u}^*, \mathbf{u}> \mathbf{v}$ because the tensor product of two vectors can be given a meaning in terms of bilinear functionals by means of this nice isomorphism:

Proposition 2.11 *Let V and U be two finite-dimensional vector spaces. Then, $L(U^*, V)$ and $L(V^* \times U^*, V)$ are* canonically isomorphic *, that is, there exists an isomorphism between them that is independent of the choice of the bases.*

We have an alternative definition for the tensor product of two vectors:

Definition 2.12 Let V and U be two finite-dimensional vector spaces. The *tensor product* $\mathbf{v} \otimes \mathbf{u}$ *of two vectors* $\mathbf{v} \in V$ *and* $\mathbf{u} \in U$ is the bilinear functional $\mathbf{v} \otimes \mathbf{u} \in L(V^* \times U^*, \mathbb{R})$ defined by

$$\mathbf{v} \otimes \mathbf{u}\ (\mathbf{v}^*, \mathbf{u}^*) = \mathbf{v}^*(\mathbf{v})\mathbf{u}^*(\mathbf{u}), \quad \forall\ (\mathbf{v}^*, \mathbf{u}^*) \in V^* \times U^*. \tag{2.22}$$

Moreover, the space $L(V^* \times U^*, \mathbb{R})$ of all bilinear functionals over $V^* \times U^*$ is called *the tensor product of the spaces* V *and* U and is denoted as $V \otimes U$.

Example 2.7 Let $v = (-1, 0, 1, 0)$ and $u = (1, 1, 0, 0)$ be two covariant vectors in $\mathbb{R}^4$. Show that $\mathbf{v} \otimes \mathbf{u} \neq \mathbf{u} \otimes \mathbf{v}$.

Solution: We remind the reader that $(\mathbb{R}^4)^*$ is isomorphic to $\mathbb{R}^4$, so that we can identify them. Now, for any two arbitrary vectors $\mathbf{u}^* = (a, b, c, d)$ and $\mathbf{v}^* = (e, f, g, h)$ in $(\mathbb{R}^4)^*$, we have

$$\mathbf{v} \otimes \mathbf{u}\ (\mathbf{u}^*, \mathbf{v}^*) = < \mathbf{v}, \mathbf{u}^* > . < \mathbf{u}, \mathbf{v}^* >= (c - a)(e + f),$$
$$\mathbf{u} \otimes \mathbf{v}\ (\mathbf{u}^*, \mathbf{v}^*) = < \mathbf{u}, \mathbf{u}^* > . < \mathbf{v}, \mathbf{v}^* >= (a + b)(g - e).$$

The next step is to characterize a basis for $V \otimes U$ from the bases of the individual spaces involved.

Proposition 2.12

Let $\{\mathbf{e}_i\}$ and $\{\mathbf{e}^{*i}\}$, $i = 1, 2, \ldots, m$, be the bases of V and V^*, respectively. Let $\{\mathbf{f}_j\}$ and $\{\mathbf{f}^{*j}\}$, $j = 1, 2, \ldots, n$, be the bases of U and U^*, respectively. Then, the set $\{\mathbf{e}_i \otimes \mathbf{f}_j;\ i = 1, \ldots, m$ and $j = 1, \ldots, n\}$ is a basis for the tensor product space $V \otimes U$.

The basis $\{\mathbf{e}_i \otimes \mathbf{f}_j;\ i = 1, \ldots, m$ and $j = 1, \ldots, n\}$ from the proposition above is called *the product basis* for $V \otimes U$ and the coefficients T^{ij} are called the *components* of $\mathbf{T}$ with respect to this basis. Moreover,

$$\dim L(V^* \times U^*, \mathbb{R}) = \dim V \otimes U = m \times n = \dim L(U^*, V)$$

so that $L(V^* \times U^*, \mathbb{R})$ and $L(U^*, V)$ have the same dimension.

In order to have the components of $\mathbf{T}$ for the basis $\mathbf{g}_k \otimes \mathbf{h}_i$, consider two n-dimensional vector spaces V and U and take $T \in L(V, U)$. Let $\{\mathbf{g}_1, \ldots, \mathbf{g}_n\}$ and $\{\mathbf{h}_1, \ldots, \mathbf{h}_n\}$ be holonomics bases for V and U, respectively. Since $T(\mathbf{g}_k) \in U$, we can write

$$T(\mathbf{g}_k) = T^i_k \mathbf{h}_i,$$

where

$$T^i_k = [T^i_k]^{\mathbf{f}}_{\mathbf{e}}.$$

Now, let $\{\mathbf{h}^{*1}, \ldots, \mathbf{h}^{*n}\}$ be the dual basis for $\{\mathbf{h}_1, \ldots, \mathbf{h}_n\}$. Applying $\mathbf{h}^{*j}$ to both sides of (2.16), we have

$$\mathbf{h}^{*j}(T(\mathbf{e}_k)) = T^i_k \mathbf{h}^{*j}(\mathbf{h}_i) = T^i_k \delta^j_i = T^j_k,$$

as implied by the definition of dual basis. Therefore, $T_k^j = \mathbf{h}^{*j}(T(\mathbf{g}_k))$ are the components of $\mathbf{T}$ with respect to the bases $\{\mathbf{h}_1, \ldots, \mathbf{h}_n\}$ and $\{\mathbf{g}^{*1}, \ldots, \mathbf{g}^{*n}\}$. The $m \times n$ scalars T_k^i are called *the components of tensor* $\mathbf{T}$ for the product basis $\mathbf{g}^{*k} \otimes \mathbf{h}_i$.

In order to distinguish such components T_k^i from other components of $\mathbf{T}$, we can explicitly indicate the bases by using brackets, that is,

$$T_k^i = [T_k^i]_{\mathbf{g}}^{\mathbf{h}} = [T_k^i]_{\mathbf{g}^* \otimes \mathbf{h}}$$

and

$$\mathbf{T} = T_k^i \mathbf{g}^{\mathbf{k}} \otimes \mathbf{h}_i \quad \text{or} \quad \mathbf{T}(\mathbf{g}_k) = T_k^i \mathbf{h}_i.$$

Proposition 2.13 *Let* $\mathbf{e} = \{\mathbf{e}_1, \mathbf{e}_2, ..., \mathbf{e}_n\}$ *and* $\mathbf{f} = \{\mathbf{f}_1, \mathbf{f}_2, ..., \mathbf{f}_n\}$ *be two arbitrary bases on* V. *Then, the tensors* $\mathbf{e}_i \otimes \mathbf{f}_j$, *for* $i, j = 1, 2, ..., n$, *form a basis for* $V \otimes V \sim L(V^*, V)$, *so that the dimension of the vector space* $L(V^*, V)$ *is* n^2.

If $\mathbf{T} \in L(V^*, V)$ is an arbitrary second-order tensor, we can write

$$\mathbf{T} = T^{ij} \mathbf{e}_i \otimes \mathbf{f}_j \quad \text{or} \quad \mathbf{T}(\mathbf{e}^i) = T^{ij} \mathbf{f}_j$$

and for the representation of second-order tensors we can use one of the following holonomics bases $\mathbf{g}_i \otimes \mathbf{g}_j$, $\mathbf{g}^i \otimes \mathbf{g}^j$, $\mathbf{g}_i \otimes \mathbf{g}^j$, or $\mathbf{g}^i \otimes \mathbf{g}_j$ $(i, j = 1, 2, ..., n)$. With respect to these bases, a tensor $\mathbf{T}$ can be written as

$$\mathbf{T} = T^{ij} \mathbf{g}_i \otimes \mathbf{g}_j = T_{ij} \mathbf{g}^i \otimes \mathbf{g}^j = T_j^i g_i \otimes g^j = T_i^j \mathbf{g}^i \otimes \mathbf{g}_j.$$

For an arbitrary anholonomic basis $\mathbf{b}$ we will use the first letters of the alphabet a, b, c, d

$$\mathbf{T} = T^{ac} \mathbf{b}_a \otimes \mathbf{b}_c = T_{ac} \mathbf{b}^a \otimes \mathbf{b}^c = T_c^a \mathbf{b}_a \otimes \mathbf{b}^c = T_a^c \mathbf{b}^a \otimes \mathbf{b}_c.$$

In the following chapters we will use also special anholonomic bases indicated by $\mathbf{e}$ and $\mathbf{f}$. In this case, a tensor of second-order may be represented by various combinations of these two bases. For example,

$$\mathbf{T} = T^{ac} \mathbf{e}_a \otimes \mathbf{e}_c = T_{ac} \mathbf{e}^a \otimes \mathbf{e}^c = T_c^a \mathbf{e}_a \otimes \mathbf{e}^c = T_a^c \mathbf{e}^a \otimes \mathbf{e}_c,$$

or

$$\mathbf{T} = T^{ac} \mathbf{f}_a \otimes \mathbf{f}_c = T_{ac} \mathbf{f}^a \otimes \mathbf{f}^c = T_c^a \mathbf{f}_a \otimes \mathbf{f}^c = T_a^c \mathbf{f}^a \otimes \mathbf{f}_c,$$

or yet

$$\mathbf{T} = T^{ac} \mathbf{e}_a \otimes \mathbf{f}_c = T_{ac} \mathbf{e}^a \otimes \mathbf{f}^c = T_c^a \mathbf{e}_a \otimes \mathbf{f}^c = T_a^c \mathbf{e}^a \otimes \mathbf{f}_c.$$

We shall see later that the introduction of this pair of bases yields that there are 24 possible representations for a second-order tensor $\mathbf{T}$.

Example 2.8 Obtain spaces that are isomorphic to

$$(a)V^* \otimes U^*, \quad (b)V \otimes U^*, \quad (c)V^* \otimes U$$

and define the corresponding tensor products.

Solution:
(a) We can write

$$\begin{aligned} V^* \otimes U^* &\sim L(V \times U, \mathbb{R}) \sim L(U, V^*) \\ (\mathbf{v}^* \otimes \mathbf{u}^*)(\mathbf{v}, \mathbf{u}) &= <\mathbf{v}^*, \mathbf{v}><\mathbf{u}^*, \mathbf{u}> \\ (\mathbf{v}^* \otimes \mathbf{u}^*)(\mathbf{u}) &= <\mathbf{u}^*, \mathbf{u}> \mathbf{v}^* \\ \mathbf{T} &= T_{ij}\mathbf{e}^i \otimes \mathbf{f}^j, \quad \forall \mathbf{T} \in V^* \otimes U^*, \end{aligned}$$

where $\{\mathbf{e}^i \otimes \mathbf{f}^j;\ i = 1, \ldots, m;$ and $j = 1, \ldots, n\}$ is a basis for $V^* \otimes U^*$. Recall that $\sim$ denotes that the spaces are isomorphic.
(b)

$$\begin{aligned} V \otimes U^* &\sim L(V^* \times U, \mathbb{R}) \sim L(U, V) \\ (\mathbf{v} \otimes \mathbf{u}^*)(\mathbf{v}^*, \mathbf{u}) &= <\mathbf{v}^*, \mathbf{v}><\mathbf{u}^*, \mathbf{u}> \\ (\mathbf{v} \otimes \mathbf{u}^*)(\mathbf{u}) &= <\mathbf{u}^*, \mathbf{u}> \mathbf{v} \\ \mathbf{T} &= T^i_j\mathbf{e}_i \otimes \mathbf{f}^j, \quad \forall \mathbf{T} \in V \otimes U^*, \end{aligned}$$

where $\{\mathbf{e}_i \otimes \mathbf{f}^j;\ i = 1, \ldots, m$ and $j = 1, \ldots, n\}$ is a basis for $V \otimes U^*$.
(c)

$$\begin{aligned} V^* \otimes U &\equiv L(V \times U^*, \mathbb{R}) \equiv L(U^*, V^*) \\ (\mathbf{v}^* \otimes \mathbf{u})(\mathbf{v}, \mathbf{u}^*) &= <\mathbf{v}^*, \mathbf{v}><\mathbf{u}^*, \mathbf{u}> \\ (\mathbf{v}^* \otimes \mathbf{u})(\mathbf{u}^*) &= <\mathbf{u}^*, \mathbf{u}> \mathbf{v}^* \\ \mathbf{T} &= T^j_i\mathbf{e}^i \otimes \mathbf{f}_j, \quad \forall \mathbf{T} \in V^* \otimes U, \end{aligned}$$

where $\{\mathbf{e}^i \otimes \mathbf{f}_j;\ i = 1, \ldots, m$ and $j = 1, \ldots, n\}$ is a basis for $V^* \otimes U$.

We can generalize the concept of tensor products to more than two vector spaces. In order to do this, let $V_1, V_2, \ldots, V_n$ be vector spaces. A *multilinear functional* is a mapping $\mathbf{T} : V_1 \times V_2 \times \cdots \times V_n \longrightarrow \mathbb{R}$ that is linear in each of its variable whilst the others are held constant. We denote $L(V_1 \times V_2 \times \cdots \times V_n, \mathbb{R})$ the set of all multilinear functionals defined on $V_1 \times V_2 \times \cdots \times V_n$ with the vector space structure given by the obvious addition and multiplication by scalar.

Definition 2.13 The vector space $L(V_1^* \times V_2^* \times \cdots \times V_n^*, \mathbb{R})$ is called the *tensor product space of* $V_1, V_2, \ldots, V_n$ and we denote

$$L(V_1^* \times V_2^* \times \cdots \times V_n^*, \mathbb{R}) \triangleq V_1 \otimes V_2 \otimes \cdots \otimes V_n.$$

We call the *tensor product* $\mathbf{v}_1 \otimes \mathbf{v}_2 \otimes \cdots \otimes \mathbf{v}_n$ of n vectors $\mathbf{v}_i \in V_i,\ i = 1, 2, \ldots, n$ the mapping defined by

$$(\mathbf{v}_1 \otimes \mathbf{v}_2 \otimes \cdots \otimes \mathbf{v}_n)(\mathbf{v}_1^*, \mathbf{v}_2^*, \ldots, \mathbf{v}_n^*) :=< \mathbf{v}_1^*, \mathbf{v}_1 >< \mathbf{v}_2^*, \mathbf{v}_2 > \cdots < \mathbf{v}_n^*, \mathbf{v}_n >,$$

for all $\mathbf{v}_i^* \in V_i^*,\ i = 1, 2, \ldots, n.$

2.2.5 *Transformation of tensor components due to change of bases*

Let $\mathbf{e} = \{\mathbf{e}_1, \mathbf{e}_2, \ldots, \mathbf{e}_n\}$ and $\mathbf{f} = \{\mathbf{f}_1, \mathbf{f}_2, \ldots, \mathbf{f}_n\}$ be two arbitrary bases of V, related by

$$\mathbf{f}_{i'} = P_{i'}^{j} \mathbf{e}_j, \quad \mathbf{e}_i = Q_i^{j'} \mathbf{f}_{j'} \tag{2.23}$$

and

$$\mathbf{f}^{*j'} = Q_i^{j'} \mathbf{e}^{*i}, \quad \mathbf{e}^{*j} = P_{i'}^{j} \mathbf{e}^{*i'}. \tag{2.24}$$

Here, the coefficients satisfy the relations

$$P_{k'}^{j} Q_k^{k'} = \delta_k^j \quad \text{and} \quad Q_j^{i'} P_k^j = \delta_k^{i'}.$$

On the other hand, the component forms relative to the bases $\mathbf{f}$ and $\mathbf{e}$ are

$$\begin{aligned} \mathbf{T} &= T_{j'_1 \cdots j'_q}^{i'_1 \cdots i'_p} \mathbf{f}_{i'_1} \otimes \cdots \otimes \mathbf{f}_{i'_p} \otimes \mathbf{f}^{*j'_1} \otimes \cdots \otimes \mathbf{f}^{*j'_q} \\ &= T_{j_1 \cdots j_q}^{i_1 \cdots i_p} \mathbf{e}_{i_1} \otimes \cdots \otimes \mathbf{e}_{i_p} \otimes \mathbf{e}^{*j_1} \otimes \cdots \otimes \mathbf{e}^{*j_q} \end{aligned} \tag{2.25}$$

Equating the two component forms in (2.25) and taking into account the formulas of transformation (2.23) and (2.24), we obtain

$$\begin{aligned} &T_{j'_1 \cdots j'_q}^{i'_1 \cdots i'_p} \mathbf{f}_{i'_1} \otimes \cdots \otimes \mathbf{f}_{i'_p} \otimes \mathbf{f}^{*j'_1} \otimes \cdots \otimes \mathbf{f}^{*j'_q} \\ &\quad = T_{j_1 \cdots j_q}^{i_1 \cdots i_p} \mathbf{e}_{i_1} \otimes \cdots \otimes \mathbf{e}_{i_p} \otimes \mathbf{e}^{*j_1} \otimes \cdots \otimes \mathbf{e}^{*j_q} \\ &\quad = T_{j_1 \cdots j_q}^{i_1 \cdots i_p} Q_{i_1}^{i'_1} \cdots Q_{i_p}^{i'_p} P_{j'_1}^{j_1} \cdots P_{j'_q}^{j_q} \mathbf{e}_{i'_1} \otimes \cdots \otimes \mathbf{e}_{i'_p} \otimes \mathbf{e}^{*j'_1} \otimes \cdots \otimes \mathbf{e}^{*j'_q}. \end{aligned}$$

Therefore,

$$T^{i'_1 \cdots i'_p}_{j'_1 \cdots j'_q} = Q^{i'_1}_{i_1} \cdots Q^{i'_p}_{i_p} P^{j_1}_{j'_1} \cdots P^{j_q}_{j'_q} T^{i_1 \cdots i_p}_{j_1 \cdots j_q} \tag{2.26}$$

and

$$T^{i_1 \cdots i_p}_{j_1 \cdots j_q} = P^{i_1}_{i'_1} \cdots P^{i_p}_{i'_p} Q^{j'_1}_{j_1} \cdots Q^{j'_q}_{j_q} T^{i'_1 \cdots i'_p}_{j'_1 \cdots j'_q}, \tag{2.27}$$

which are the transformation formulas for the components of any tensor $\mathbf{T} \in T^p_q(V)$. In particular, for a second-order tensor, (2.26) and (2.27) take the form

$$T^{i'}_{j'} = T^i_j Q^{i'}_i P^j_{j'}, \quad T^i_j = T^{i'}_{j'} P^i_{i'} Q^{j'}_j,$$

respectively.

Now, consider two frames $\mathbf{b} = \{\mathbf{b}_a\}$ and $\mathbf{g} = \{\mathbf{g}_k\}$ and their dual $\mathbf{b}^* = \left\{\mathbf{b}^{*a}\right\}$ and $\mathbf{g}^* = \left\{\mathbf{g}^{*k}\right\}$. Suppose also that such frames are related by means of two matrices J and N, that is,

$$\mathbf{b} = J\mathbf{g} \tag{2.28}$$

and

$$\mathbf{b}^* = N\mathbf{g}^*. \tag{2.29}$$

In terms of their components, these formulas yield

$$\mathbf{b}_a = J_a^{\ k}.\mathbf{g}_k \quad \text{and} \quad \mathbf{b}^{*\,a} = N^a_{k}.\ \mathbf{g}^{*\,k}. \tag{2.30}$$

We claim that the matrices J and N are related by

$$N^T = J^{-1} \iff N_k^{\ a} = \left(J^{-1}\right)_k^{\ a}.$$

Indeed, take the transpose of (2.29), $\mathbf{b}^{*\,T} = \mathbf{g}^{*\,T} N^T$. Then,

$$\mathbf{b}\,\mathbf{b}^{*\,T} = J\mathbf{g}\,\mathbf{g}^{*\,T} N^T. \tag{2.31}$$

But $\mathbf{b}\mathbf{b}^{*T} = \mathbf{g}\mathbf{g}^{*T} = I$ (identity matrix), so that

$$I = JN^T \iff J_a^{\ k} N_k^{\ p} = \delta_a^{\ p},$$

or yet

$$I^T = N.J^T \iff N^p_{\ k}\, J^k_{\ a} = \delta^p_{\ a}.$$

Similarly, we have $N_s^{\ a} J_a^{\ k} = \delta_s^{\ k}$, and so on, which proves that $N^T = J^{-1}$.

2.2.6 *Change of bases for tensors*

Now, we can see how the change of a basis ${\beta_{ab}}^c = \mathbf{b}_a \otimes \mathbf{b}_b \otimes \mathbf{b}^{*c}$ is formed by a tensor product of vectors. Using the formula (2.29) we can see that

$${\beta_{ab}}^c = \mathbf{b}_a \otimes \mathbf{b}_b \otimes \mathbf{b}^{*\,c} = {J_a}^k {J_b}^j N^c_i \mathbf{g}_k \otimes \mathbf{g}_j \otimes \mathbf{g}^{*\,i} = {J_a}^k {J_b}^j N^c_i {\gamma_{kj}}^i$$

where ${\gamma_{kj}}^i = \mathbf{g}_k \otimes \mathbf{g}_j \otimes \mathbf{g}^{*\,i}$. Summarizing this result, we can write

$${\beta_{ab}}^c = {J_a}^k {J_b}^j N^c_i \left({\gamma_{kj}}^i\right).$$

If we multiply both members of the above formula by the inverses of J and N, we obtain

$${\gamma_{kj}}^i = {N_k}^a {N_j}^b {J^i}_c \left({\beta_{ab}}^c\right).$$

Thus, if $T = T^{ab}_{\dots c} \mathbf{b}_a \otimes \mathbf{b}_b \otimes \mathbf{b}^{*\,c} = T^{kj}_{\dots i} \mathbf{g}_k \otimes \mathbf{g}_j \otimes \mathbf{g}^{*\,i}$, then

$$T^{ab}_{\dots c} {J_a}^k {J_b}^j N^c_i \mathbf{g}_k \otimes \mathbf{g}_j \otimes \mathbf{g}^{*\,i} = T^{kj}_{\dots i} \mathbf{g}_k \otimes \mathbf{g}_j \otimes \mathbf{g}^{*\,i},$$

which implies the following transformation of components

$$T^{ab}_{\dots c} = J^a_k {J_b}^j N^c_i = T^{kj}_{\dots i}.$$

Similarly,

$$T^{kj}_{\dots i} {N_k}^a {N_j}^b {J^i}_c = T^{ab}_{\dots c}.$$

In the case of tensors of second-order, we obtain the following formulas

$$\begin{aligned} T^{ab} &= T^{kj} {N_k}^a {N_j}^b, \\ {T^a}_c &= T^k_{.i} {N_k}^a {J^i}_c, \\ T_{ac} &= T_{ki} {J^k}_a {J^i}_c. \end{aligned}$$

Generalizing to a tensor $\mathbf{T}$ of order n, these formulas read

$$T^{abc\dots}_{prq\dots} = {N_i}^a {N_j}^b {N_k}^c \dots {J^s}_p {J^t}_q {J^u}_r T^{ijk..}_{siu..},$$

where $a, b, c..., p, q, r$ are indexes of an anholonomic basis and s, t, u, i, j, k are indexes of a natural basis.

Now, consider $\mathbf{v}_1, \dots, \mathbf{v}_p \in V$ and $\mathbf{v}^{*1}, \dots, \mathbf{v}^{*q} \in V^*$. An example of a tensor in $T^p_q(V)$ is the tensor product of these vectors, that is, $\mathbf{T} = \mathbf{v}_1 \otimes \dots \otimes \mathbf{v}_p \otimes \mathbf{v}^{*1} \otimes \dots \otimes \mathbf{v}^{*q}$

$$\text{from } \underbrace{V^* \times V^* \times \dots \times V^*}_{p \text{ times}} \times \underbrace{V \times V \times \dots \times V}_{q \text{ times}} \text{ into } \mathbb{R}$$

defined by setting $\mathbf{T}(\mathbf{u}^{*1}, \dots, \mathbf{u}^{*p}, \mathbf{u}_1, \dots, \mathbf{u}_q)$ equal to

$$< \mathbf{u}^{*1}, \mathbf{v}_1 > \cdots < \mathbf{u}^{*p}, \mathbf{v}_p >< \mathbf{v}^{*1}, \mathbf{u}_1 > \cdots < \mathbf{v}^{*q}, \mathbf{u}_q >,$$

for all $\mathbf{u}_1, \dots, \mathbf{u}_q \in V$ and $\mathbf{u}^{*1}, \dots, \mathbf{u}^{*p} \in V^*$.

A straightforward generalization of Proposition 2.13 is the following:

Proposition 2.14 *Let* $\{\mathbf{e}_i\}$ *and* $\{\mathbf{e}^{*i}\}$, $i = 1, \dots, n$, *be dual bases of* V *and* V^*, *respectively. Then, the set* $\{\mathbf{e}_{i_1} \otimes \cdots \otimes \mathbf{e}_{i_p} \otimes \mathbf{e}^{*j_1} \otimes \cdots \otimes \mathbf{e}^{*j_q}\}$ *is a basis for* $T^p_q(V)$, *called the*product basis. *Hence,*

$$\dim T^p_q(V) = n^{p+q}.$$

Thus, a tensor $\mathbf{T} \in T^p_q(V)$ has the following component representation:

$$\mathbf{T} = T^{i_1 \cdots i_p}_{j_1 \cdots j_q} \mathbf{e}_{i_1} \otimes \cdots \otimes \mathbf{e}_{i_p} \otimes \mathbf{e}^{*j_1} \otimes \cdots \otimes \mathbf{e}^{*j_q}. \tag{2.32}$$

In particular, for a second-order tensor, (2.26) and (2.27) take the following respective forms:

$$T^{i'}_{j'} = T^i_j Q^{i'}_i P^j_{j'}, \tag{2.33}$$

and

$$T^i_j = T^{i'}_{j'} P^i_{i'} Q^{j'}_j. \tag{2.34}$$

2.2.7 *Algebraic operations for components of tensors*

Addition, tensorial multiplication, self-contraction, product contraction, symmetrization, and alternation are the main operations when one deals with algebraic aspects of components of tensors. Here, we have some words to say about each of these operations.

1. Addition. It applies only to tensors of the same order and type. If we are given two systems of components of tensors of the same order and type, and if we add each component of the first tensor to the corresponding component of the second, we obviously arrive at a set of components of a tensor of the same order and type as the original components. This process is the operation of addition and the resulting components are called the *sum* of the two components.Thus, if S^{ab}_{cde} and P^{ab}_{cde} are two sets of components of tensors of order 5, then the components of the sum $\mathbf{T} = \mathbf{S} + \mathbf{P}$ of $\mathbf{S}$ and $\mathbf{P}$ are defined by the equations

$$T^{ab}_{cde} = S^{ab}_{cde} + P^{ab}_{cde}. \tag{2.35}$$

2. Tensorial multiplication. If we take two systems of components of tensors of any kind and multiply each component of the first by each component of the second, we get a system of components of tensors, called their product, whose order equals the sum of the orders of the two original systems. For example, if S^{ab}_{cd} are the components of a tensor $\mathbf{S}$ and P_{ef} are the components of a tensor $\mathbf{P}$, then their product is the tensor $\mathbf{T}$ with components given by

$$T^{ab}_{cdef} = S^{ab}_{cd} P_{ef}. \tag{2.36}$$

This process can be extended to define the product of any number of systems of components of tensors.

3. Self-contraction. This operation allows one to create new tensors of lower order for any given tensor. Such process can be better explained by means of an example: Let us take a system of components of tensor of order k, say $T^{a}_{.bc...k}$, which has both upper and lower indices. If we put $a = b$, we get the set $T^{b}_{bc...k}$. Since b is now a repeated index, it must be summed from 1 to n, in accordance to our sum convention. Thus, the new system obtained in this way is $S_{c...k} = T^{b}_{bc...k}$ which defines a tensor $\mathbf{S}$ of order $k - 2$. Of course, the operation can be repeated several times because we can contract with respect to any pair of indices, one being a subscript and the other, a superscript.

4. Product contraction. For this operation, we first take the product of two systems of components of tensors and then contract the result with respect to an index, subscript of one and a superscript of the other. These operations can be describe in terms of tensor in the following way:

$$\begin{aligned}\mathbf{A} &= \mathbf{P} \cdot \mathbf{S} \\ &= (P^{abd}\mathbf{e}_a \otimes \mathbf{e}_b \otimes \mathbf{e}_d) \cdot (S_{cd}\mathbf{e}^c \otimes \mathbf{e}^d) \\ &= P^{abd} S_{cd}\mathbf{e}_a \otimes \mathbf{e}_b \otimes \mathbf{e}^c(\mathbf{e}_d \cdot \mathbf{e}^d) \\ &= P^{abd} S_{cd}\mathbf{e}_a \otimes \mathbf{e}_b \otimes \mathbf{e}^c.\end{aligned}$$

Since $\mathbf{A} = A^{ab}_{c}\mathbf{e}_a \otimes \mathbf{e}_b \otimes \mathbf{e}^c$, it follows that

$$P^{abd} S_{cd} = A^{ab}_{c}.$$

5. Symmetrization. This process is applied always to a number of upper and lower indices. In order to perform such an operation over k indices, we form $k!$ scalars by permuting these indices in all possible ways and then we average by taking the sum of these scalars and divide by $k!$. The operation of symmetrization is denoted **Sym**. For example,

$$\mathbf{Sym}(T_{ab}) = \frac{T_{ab} + T_{ba}}{2!},$$

$$\mathbf{Sym}(T_{abc}) = \frac{T_{abc} + T_{bca} + T_{cab} + T_{acb} + T_{cba} + T_{bac}}{3!}.$$

6. Alternation. This operation, denoted either by **Alt** or by a pair of square brackets [ ], is perfomed in the same way as the process of symmetrization, only that the averaging is taken with the positive sign if the permutation is even and the negative sign if the permutation is odd. Here are two examples:

$$\mathbf{Alt}(T_{abc}) = T_{[abc]} = \frac{T_{abc} + T_{bca} + T_{cab} - T_{acb} - T_{cba} - T_{bac}}{3!},$$

$$\mathbf{Alt}(T_{ab}) = T_{[ab]} = \frac{T_{ab} - T_{ba}}{2!}.$$

2.3 Exercises

(1) Let $\{\mathbf{i}, \mathbf{j}, \mathbf{k}\}$ be a right-handed orthonormal basis. Show that, for

$$\mathbf{A} = A_1\mathbf{i} + A_2\mathbf{j} + A_3\mathbf{k}, \quad \mathbf{B} = B_1\mathbf{i} + B_2\mathbf{j} + B_3\mathbf{k},$$

we have that $\mathbf{A} \times \mathbf{B} = C_1\mathbf{i} + C_2\mathbf{j} + C_3\mathbf{k}$, where

$$\begin{Bmatrix} C_1 \\ C_2 \\ C_3 \end{Bmatrix} = \begin{bmatrix} 0 & -A_3 & A_2 \\ A_3 & 0 & -A_1 \\ -A_2 & A_1 & 0 \end{bmatrix} \begin{Bmatrix} B_1 \\ B_2 \\ B_3 \end{Bmatrix}.$$

(2) Show that

$$(\mathbf{a} \cdot \mathbf{b} \times \mathbf{c})(\mathbf{d} \cdot \mathbf{e} \times \mathbf{f}) = \begin{bmatrix} \mathbf{a} \cdot \mathbf{d} & \mathbf{a} \cdot \mathbf{e} & \mathbf{a} \cdot \mathbf{f} \\ \mathbf{b} \cdot \mathbf{d} & \mathbf{b} \cdot \mathbf{e} & \mathbf{b} \cdot \mathbf{f} \\ \mathbf{c} \cdot \mathbf{d} & \mathbf{c} \cdot \mathbf{e} & \mathbf{c} \cdot \mathbf{f} \end{bmatrix}.$$

(3) Compute the volume element dV corresponding to the prolate spheroidal coordinates (u, v, ϕ) given by

$$\begin{aligned} x =& a \sinh y \sin v \cos \phi, \\ y =& a \sinh u \sin v \sin \phi, \\ z =& a \cosh u \cos v. \end{aligned}$$

Answer: $dV = \left(a^2 \sqrt{\sinh^2 u + \sin^2 v}\right) \sinh u \sin v \, du dv d\phi.$

(4) Let $\mathbf{T} \in V \otimes W$, $\mathbf{T} \neq \mathbf{0}$. Show that in general there exist several representations of the form

$$\mathbf{T} = \mathbf{v}_i \otimes \mathbf{w}_i, \qquad \mathbf{v}_i \in V, \ \mathbf{w}_i \in W,$$

with $\{\mathbf{v}_1, \ldots, \mathbf{v}_n\}$ and $\{\mathbf{w}_1, \ldots, \mathbf{w}_n\}$ each linearly independent.

(5) Let $\mathbf{a} \in V$ and $\mathbf{b} \in W$ be such that $\mathbf{b} \otimes \mathbf{a} \neq \mathbf{0}$. Prove that

$$\mathbf{b} \otimes \mathbf{a} = \mathbf{w} \otimes \mathbf{v} \quad \text{if and only if} \quad \mathbf{v} = \lambda \mathbf{a} \text{ and } \mathbf{w} = \lambda \mathbf{b}, \text{ for some } \lambda \neq 0.$$

(6) Show that a self-contraction over two contravariant or two covariant indices of a tensor does not give another tensor. For example, T_{abc} is not a tensor.

(7) Show that $L(V, L(V, V))$ is isomorphic to $T_2^1(V)$.

(8) Prove that $g_{ij}\mathbf{A}^i\mathbf{B}^j$ is invariant, i.e., $g_{\overline{ij}}\mathbf{A}^{\overline{i}}\mathbf{B}^{\overline{j}} = g_{ij}\mathbf{A}^i\mathbf{B}^j$.

(9) Associated to vectors $\mathbf{a} = a_i\mathbf{e}^i$, $\mathbf{b} = b_j\mathbf{e}^j$, and $\mathbf{c} = c_k\mathbf{e}^k$ it is possible to define the following trilinear functional $\mathbf{T}$ by putting

$$\mathbf{T}(\mathbf{a}, \mathbf{b}, \mathbf{c}) = a_i b_j c_k T^{ij}_{k}.$$

Show that this functional defines a tensor.

(10) Associated to the standard basis $\{\mathbf{e}_1, \mathbf{e}_2, \mathbf{e}_3\}$ of $\mathbb{R}^3$,

$$\mathbf{e}_1 = \begin{Bmatrix} 1 \\ 0 \\ 0 \end{Bmatrix}, \quad \mathbf{e}_2 = \begin{Bmatrix} 0 \\ 1 \\ 0 \end{Bmatrix}, \quad \mathbf{e}_3 = \begin{Bmatrix} 0 \\ 0 \\ 1 \end{Bmatrix}$$

there exists a set of scalars defined by

$$T^i_{jk} = T(\mathbf{e}^i, \mathbf{e}_j, \mathbf{e}_k)$$

where T is a given. Suppose another basis is defined as

$$\mathbf{e}_{\overline{1}} = \mathbf{e}_1 - 2\mathbf{e}_2, \quad \mathbf{e}_{\overline{2}} = 2\mathbf{e}_1 + \mathbf{e}_2, \quad \mathbf{e}_{\overline{3}} = \mathbf{e}_1 + \mathbf{e}_2.$$

Find $T^{\overline{i}}_{\overline{jk}} = T(\mathbf{e}^{\overline{i}}, \mathbf{e}_{\overline{j}}, \mathbf{e}_{\overline{k}})$ in terms of T^i_{jk}.

(11) Prove that $T^i_j\mathbf{v}_i\mathbf{w}^j$ is invariant.

(12) Prove that if $a_j = T_{ij}\mathbf{v}^i$ are the components if a covariant vector $\mathbf{a}$ for all contravariant vectors $\mathbf{v}$, then T_{ij} is a covariant tensor of order 2.

(13) If v^i are the components of a contravariant vector, when can you say that a^i defined by $a^i = v^i + \alpha v_i$, $\alpha \in \mathbb{R}$, are components of a vector $\mathbf{a}$?

Chapter 3

Tensor Calculus

The tensor calculus extends the notion of derivative of scalar quantities and vectors to more complex entities called tensors by combining methods from linear algebra and vector calculus. Basically, the required tools are the changing of bases of vector spaces and the transformation of coordinates between two different systems of coordinates.

3.1 Tensor fields

3.1.1 *Gradient of a field*

Consider a map $f : \Omega \to W$, where Ω is an open subset of a Euclidian space V and W is a normed vector space. In the case where W is

$$\mathbb{R} \quad \text{or} \quad V \quad \text{or} \quad L\left(\underbrace{V \times V \times \cdots \times V}_{n}, \mathbb{R}\right),$$

we say that f is, respectively, a scalar field, a vector field, or a tensor field of order n on Ω. Examples of scalar, vector, and tensor fields are, in that order, the temperature at any point in a solid, the velocity at any point of a fluid, and the strain at any point of a deformable solid.

Recall that we say that f is *differentiable* at $\mathbf{x} \in \Omega$ iff there exist a linear transformation $Df(\mathbf{x}) \in L(V, W)$ and a map $\mathcal{O}(.) : V \to W$ such that, for all $\mathbf{v} \in V$ with $\mathbf{x} + \mathbf{v} \in \Omega$,

$$f(\mathbf{x} + \mathbf{v}) = f(\mathbf{x}) + Df(\mathbf{x})\ (\mathbf{v}) + \mathcal{O}(\mathbf{v}) \tag{3.1}$$

with

$$\lim_{||\mathbf{v}|| \to 0} \frac{\mathcal{O}(\mathbf{v})}{||\mathbf{v}||} = \mathbf{0}. \tag{3.2}$$

Usually, we take $V = \mathbb{R}^n$ and $\mathbf{x}$ will stand for the position vector of an arbitrary point in Ω relative to the origin O. Note that, since $\mathcal{D}$ is open, there exists a $\delta > 0$ such that $||\mathbf{v}|| < \delta \Longrightarrow \mathbf{x} + \mathbf{v} \in \Omega$. .

Theorem 3.1 *If f is differentiable at $\mathbf{x}$, then we have that $Df(\mathbf{x})$ is uniquely determined for each $\mathbf{x}$.*

■ **Proof:** Suppose $Df_1(\mathbf{x}),\ Df_2(\mathbf{x}) \in L(V, W)$ are such that

$$f(\mathbf{x}+\mathbf{v}) = f(\mathbf{x}) + Df_1(\mathbf{x})\ (\mathbf{v}) + \mathcal{O}_1(\mathbf{v}),$$
$$f(\mathbf{x}+\mathbf{v}) = f(\mathbf{x}) + Df_2(\mathbf{x})\ (\mathbf{v}) + \mathcal{O}_2(\mathbf{v}),$$

hold with

$$\lim_{||\mathbf{v}||\to 0} \frac{\mathcal{O}_1(\mathbf{v})}{||\mathbf{v}||} = \lim_{||\mathbf{v}||\to 0} \frac{\mathcal{O}_2(\mathbf{v})}{||\mathbf{v}||} = \mathbf{0}.$$

Then we have

$$(Df_1(\mathbf{x}) - Df_2(\mathbf{x}))\ (\mathbf{v}) = \mathcal{O}_2(\mathbf{v}) - \mathcal{O}_1(\mathbf{v}).$$

So, from (3.2) it follows that

$$\lim_{||\mathbf{v}||\to 0} (Df_1(\mathbf{x}) - Df_2(\mathbf{x})) \left(\frac{\mathbf{v}}{||\mathbf{v}||}\right) = \mathbf{0}.$$

Thus, the linearity of the transformations $Df_1(\mathbf{x})$ and $Df_2(\mathbf{x})$ yield that for any given $\varepsilon > 0$, there corresponds a $\delta > 0$ such that

$$||Df_1(\mathbf{x}) - Df_2(\mathbf{x})\ (\mathbf{v})|| < \varepsilon ||\mathbf{v}||, \qquad \text{for } ||\mathbf{v}|| < \delta$$

But this implies that $||Df_1(\mathbf{x}) - Df_2(\mathbf{x})|| < \varepsilon$ for ε arbitrary. Hence, $Df_1(\mathbf{x}) = Df_2(\mathbf{x})$. □

From (3.1), taking vectors of the form $t\mathbf{v}$ for scalar $t \neq 0$ such that $\mathbf{x} + t\mathbf{v} \in \Omega$, we can write

$$Df(\mathbf{x})\ (\mathbf{v}) = \frac{f(\mathbf{x}+t\mathbf{v}) - f(\mathbf{x})}{t} + \frac{1}{t}\mathcal{O}(t\mathbf{v}).$$

Now, since $\lim_{t\to 0} \frac{1}{t}\mathcal{O}(\mathbf{v}) = \lim_{t\to 0} ||\mathbf{v}|| \frac{\mathcal{O}(t\mathbf{v})}{||t\mathbf{v}||} = 0$, we have

$$Df(\mathbf{x})\ (\mathbf{v}) = \lim_{t\to 0} \frac{f(\mathbf{x}+t\mathbf{v}) - f(\mathbf{x})}{t} \triangleq \frac{d}{dt} f(\mathbf{x}+t\mathbf{v})\bigg|_{t=0}. \tag{3.3}$$

This limit is called the *directional derivative of f at $\mathbf{x}$ in the direction $\mathbf{v}$*. More specifically, for scalar, vector, and tensor fields, this notion reads as follows:

(a) If $f = \varphi$ denotes a differentiable scalar field on Ω, then $D\varphi(\mathbf{x})$ is a linear mapping from V into $W = \mathbb{R}$. By the representation theorem for linear forms, there exists a vector, which we denote $\boldsymbol{\nabla}\varphi(\mathbf{x})$, such that

$$D\varphi(\mathbf{x})(\mathbf{v}) = \boldsymbol{\nabla}\varphi(\mathbf{x}) \cdot \mathbf{v}.$$

Thus, (3.3) takes the form

$$\boldsymbol{\nabla}\varphi(\mathbf{x}) \cdot \mathbf{v} = \frac{d}{dt}\varphi(\mathbf{x} + t\mathbf{v})\bigg|_{t=0}. \tag{3.4}$$

The vector $\boldsymbol{\nabla}\varphi(\mathbf{x})$ is called the *gradient* of φ at $\mathbf{x}$.

(b) If $f = \mathbf{w}$ denotes a differentiable vector field on Ω, then $D\mathbf{w}(\mathbf{x})$ is a linear mapping from V into $W = V$. Thus, $\boldsymbol{\nabla}\mathbf{w}(\mathbf{x})$ is a second-order tensor at $\mathbf{x}$ and (3.3) reads

$$\boldsymbol{\nabla}\mathbf{w}(\mathbf{x})\mathbf{v} = \frac{d\mathbf{w}}{dt}\left(\mathbf{x} + t\mathbf{v}\right)\big|_{t=0}, \tag{3.5}$$

where

$$D\mathbf{w}(\mathbf{x})(\mathbf{v}) = \boldsymbol{\nabla}\mathbf{w}(\mathbf{x})\mathbf{v}. \tag{3.6}$$

(c) If $f = \mathbb{T}$ denotes a differentiable tensor field of order n on Ω, then $D\mathbb{T}(\mathbf{x})$ can be regarded as a linear transformation from V into W. Hence, $\boldsymbol{\nabla}\mathbb{T}$ is a tensor field of order $n+1$ on Ω and (3.3) becomes

$$\boldsymbol{\nabla}\mathbb{T}(\mathbf{x})\mathbf{v} = \frac{d\mathbb{T}}{dt}\left(\mathbf{x} + t\mathbf{v}\right)\big|_{t=0}, \tag{3.7}$$

and we denote with

$$D\mathbb{T}(\mathbf{x})(\mathbf{v}) = \boldsymbol{\nabla}\mathbb{T}(\mathbf{x})\mathbf{v}. \tag{3.8}$$

We say that

- f is of class C^1 if it is differentiable at each point of Ω and *grad* f is continuous on Ω.
- f is of class C^2 if *grad* f exists and is of class C^1 on Ω, and so forth.
- f is smooth if it is of class C^r on Ω, for some $r \geq 1$.

Example 3.1 Let $\mathbf{x} \in \mathbb{R}^3$ and $f(\mathbf{x}) = \mathbf{x} \otimes \mathbf{x}$. Prove that the function f is differentiable at x.

Solution: Developing $f(\mathbf{x}+t\mathbf{v})$, we can write

$$\begin{aligned} f(\mathbf{x}+t\mathbf{v}) &= (\mathbf{x}+t\mathbf{v})\otimes(\mathbf{x}+t\mathbf{v}) = \mathbf{x}\otimes\mathbf{x} + (\mathbf{x}\otimes t\mathbf{v} + t\mathbf{v}\otimes\mathbf{x}) + t^2\mathbf{v}\otimes\mathbf{v} \\ \nabla f(\mathbf{x})\mathbf{v} &= \frac{df}{dt}(\mathbf{x}+t\mathbf{v})|_{t=0} = (\mathbf{x}\otimes\mathbf{v} + \mathbf{v}\otimes\mathbf{x} + 2t\mathbf{v}\otimes\mathbf{v})\,|_{t=0} \\ &= 2\,\mathrm{sym}(\mathbf{x}\otimes\mathbf{v}), \end{aligned} \tag{3.9}$$

where

$$\mathrm{sym}(\mathbf{x}\otimes\mathbf{v}) = \frac{1}{2}\left(\mathbf{x}\otimes\mathbf{v} + (\mathbf{x}\otimes\mathbf{v})^{\mathsf{T}}\right).$$

Comparing (3.9) and (3.1), we have

$$Df(\mathbf{x})(\mathbf{v}) = 2\mathrm{sym}(\mathbf{x}\otimes\mathbf{v}).$$

3.1.2 *Chain rule and product rule*

Let us first deal with the chain rule for the composition of tensor functions.

Proposition 3.1 *Let Ω be an open subset of a Euclidean point space $\mathcal{E}$ and Ω_1 be an open subset of a finite-dimensional normed space V. For $\mathbf{x}\in\Omega$ (recall that we are identifying the point X with the position vector $\mathbf{x}$), suppose that g is a field with $g(\Omega)\subset\Omega_1$, differentiable at $\mathbf{x}$, and f is a field differentiable at $y = g(\mathbf{x})$. Then, the composite tensor function $T(\mathbf{x}) = f(g(\mathbf{x}))$ is differentiable at $\mathbf{x}$ and*

$$DT(\mathbf{x})(\mathbf{v}) = Df(g(\mathbf{x}))(Dg(\mathbf{x})(\mathbf{v})), \qquad \forall\,\mathbf{v}\in V. \tag{3.10}$$

■ **Proof:** Since T is a differentiable tensor field, we have that for every $\mathbf{x}$ there exists a linear transformation $DT(\mathbf{x})\in L(V;W)$ defined by (3.3):

$$DT(x)(\mathbf{v}) = \lim_{t\to 0}\frac{f(g(\mathbf{x}+t\mathbf{v})) - f(g(\mathbf{x}))}{t}.$$

Considering that g is differentiable at $\mathbf{x}$, (3.1) gives

$$DT(x)(\mathbf{v}) = \lim_{t\to 0}\frac{f(\tilde{\mathbf{x}}+t\tilde{\mathbf{v}}) - f(\tilde{\mathbf{x}})}{t},$$

where $\tilde{\mathbf{v}} = Dg(\mathbf{x})(\mathbf{v})$ and $\tilde{\mathbf{x}} = g(\mathbf{x})$. Now, since f is differentiable at $\tilde{\mathbf{x}}$, we have

$$DT(x)(\mathbf{v}) = \lim_{t\to 0}\frac{f(\tilde{\mathbf{x}}) + tDf(\tilde{\mathbf{x}})(\tilde{\mathbf{v}}) - f(\tilde{\mathbf{x}})}{t} = Df(\tilde{\mathbf{x}})(\tilde{\mathbf{v}}),$$

which gives (3.10).

When g is a function of a real variable t, that is, when $\mathcal{E} = \mathbb{R}$, and denoting the differential $Dg(t)$ by the derivative $\dot{g}(t)$, (3.10) becomes the important formula

$$\frac{d}{dt} f(g(t)) = Df(g(t))(\dot{g}(t)). \tag{3.11}$$

Note that the derivative $\dot{g}(t)$ of g at $t \in \Omega \subset \mathbb{R}$ is defined as

$$\dot{g}(t) = \lim_{\gamma \to 0} \frac{1}{\gamma} \left(g(t+\gamma) - g(t) \right),$$

or equivalently

$$g(t+\gamma) = g(t) + \gamma \dot{g}(t) + \mathcal{O}(\gamma),$$

whenever g is differentiable. This, together with (3.1), gives

$$Dg(t)(\gamma) = \gamma \dot{g}(t). \tag{3.12}$$

□

Now, we move on to the product rule. We have seen that the bilinearity is the common property for all the products that we have defined so far, such as the inner product of two vectors, the tensor product of two vectors, the action of a tensor on a vector, and so on. Therefore, a product rule can be established by considering the bilinear operation $\Pi : V_1 \times V_2 \to V_3$, where V_1, V_2, and V_3 are finite-dimensional normed spaces. If $f : \Omega \to V_1$ and $g : \Omega \to V_2$ are two functions, then the product $h = \Pi(f, g)$ is the function $h : \Omega \to V_3$ defined by

$$h(\mathbf{x}) = \Pi(f(\mathbf{x}), g(\mathbf{x})), \qquad \text{for all } \mathbf{x} \in \Omega \subset \mathcal{E}.$$

The product rule can be stated as follows.

Proposition 3.2 *Suppose that f and g are differentiable at $\mathbf{x} \in \Omega$. Then their product $\Pi(f, g)$ is differentiable at $\mathbf{x}$ and*

$$Dh(\mathbf{x})(\mathbf{v}) = \Pi(f(\mathbf{x}), Dg(\mathbf{x})(\mathbf{v})) + \Pi(Df(\mathbf{x})(\mathbf{v}), g(\mathbf{x})), \tag{3.13}$$

for all $\mathbf{v} \in V$.

■ **Proof:** For the proof we refer to [Gurtin, 1982], p.27. □

Thus, the derivative of the product $\Pi(f, g)$ can be interpreted as being the derivative of Π, holding f constant, plus the derivative of Π, holding g constant.

When $\mathcal{E} = \mathbb{R}$, (3.13) reduces to

$$\dot{h}(t) = \Pi(\dot{f}(t), g(t)) + \Pi(f(t), \dot{g}(t)). \tag{3.14}$$

In particular, if (f, g) are scalar valued, $(\mathbf{u}, \mathbf{v})$ are vector-valued, and $(\mathbb{S}, \mathbb{T})$ are tensor-valued functions, we have

$$\begin{aligned}
(f\mathbf{u})^{\cdot} &= f\dot{\mathbf{u}} + \dot{f}\mathbf{u},\\
(\mathbf{u}\cdot\mathbf{v})^{\cdot} &= \mathbf{u}\cdot\dot{\mathbf{v}} + \dot{\mathbf{u}}\cdot\mathbf{v},\\
(\mathbb{T}\mathbb{S})^{\cdot} &= \mathbb{T}\dot{\mathbb{S}} + \dot{\mathbb{T}}\mathbb{S},\\
(\mathbb{S}\mathbf{u})^{\cdot} &= \mathbb{S}\dot{\mathbf{u}} + \dot{\mathbb{S}}\mathbf{u}.
\end{aligned}$$

Example 3.2 For $\Omega \subset \mathcal{E}$ and (f, g) and $(\mathbf{u}, \mathbf{v})$ as above, show that
(a) $\boldsymbol{\nabla}(fg) = f\,\boldsymbol{\nabla}g + g\,\boldsymbol{\nabla}f$,
(b) $\boldsymbol{\nabla}(f\mathbf{u}) = f\,\boldsymbol{\nabla}\mathbf{u} + \mathbf{u}\otimes\boldsymbol{\nabla}f$, and
(c) $\boldsymbol{\nabla}(\mathbf{u}\cdot\mathbf{v}) = (\boldsymbol{\nabla}\mathbf{u})^{\mathsf{T}}\mathbf{v} + (\boldsymbol{\nabla}\mathbf{v})^{\mathsf{T}}\mathbf{u}$.

Solution: (a) By the product rule (3.13), we have, for any $\mathbf{w}\in V$,

$$\boldsymbol{\nabla}(fg)\cdot\mathbf{w} = (\boldsymbol{\nabla}f\cdot\mathbf{w})g + f(\boldsymbol{\nabla}g\cdot\mathbf{w}) = ((\boldsymbol{\nabla}f)g + f(\boldsymbol{\nabla}g))\cdot\mathbf{w}$$

and the result follows.
(b) Again, the product rule allows us to write, for any $\mathbf{w}\in V$,

$$\boldsymbol{\nabla}(f\mathbf{u})\mathbf{w} = f\,\boldsymbol{\nabla}\mathbf{u}\mathbf{w} + (\boldsymbol{\nabla}f\cdot\mathbf{w})\mathbf{u} = (f\,\boldsymbol{\nabla}\mathbf{u} + \mathbf{u}\otimes\boldsymbol{\nabla}f)\mathbf{w}.$$

(c) Given any $\mathbf{w}\in V$, we have

$$\boldsymbol{\nabla}(\mathbf{u}\cdot\mathbf{v})\mathbf{w} = (\mathbf{u}\cdot\nabla\mathbf{v})\mathbf{w} + (\mathbf{v}\cdot\boldsymbol{\nabla}\mathbf{u})\mathbf{w} = ((\boldsymbol{\nabla}\mathbf{v})^{\mathsf{T}}\mathbf{u} + (\boldsymbol{\nabla}\mathbf{u})^{\mathsf{T}}\mathbf{v})\cdot\mathbf{w}.$$

3.1.3 *Other differential operators*

The divergence of a smooth vector field $\mathbf{v}$ is the scalar field defined by

$$\operatorname{div}\mathbf{v} := tr\,\boldsymbol{\nabla}\mathbf{v}. \tag{3.15}$$

The divergence of a smooth second-order tensor field $\mathbb{S}$ is the vector field defined by the condition

$$(\operatorname{div}\mathbb{S})\cdot\mathbf{a} = \operatorname{div}(\mathbb{S}^{\mathsf{T}}\mathbf{a}), \tag{3.16}$$

for any constant vector field $\mathbf{a}$.

If $\mathbf{u}$ is a vector field, its curl is the vector field defined by

$$\operatorname{curl}\ \mathbf{u} = \frac{1}{2}\,\mathbb{E} : ((\operatorname{grad}\ \mathbf{u})^{\mathsf{T}} - \operatorname{grad}\ \mathbf{u}), \tag{3.17}$$

where $\mathbb{E}$ is the third-order permutation tensor given by

$$\mathbb{E} = \varepsilon_{ijk}\mathbf{g}^i \otimes \mathbf{g}^j \otimes \mathbf{g}^k = \varepsilon^{ijk}\mathbf{g}_i \otimes \mathbf{g}_j \otimes \mathbf{g}_k \tag{3.18}$$

and

$$\varepsilon_{ijk} = \sqrt{g}e_{ijk}, \qquad \varepsilon^{ijk} = \frac{1}{\sqrt{g}}e^{ijk} \tag{3.19}$$

are its covariant and contravariant components, respectively. Here, e_{ijk} and e^{ijk} are the permutation symbols. The symbol : in (3.17) represents a *double scalar product* of two tensor products of vectors, which is defined as follows:

$$(\mathbf{a} \otimes \mathbf{b}) : (\mathbf{c} \otimes \mathbf{d}) = (\mathbf{a} \cdot \mathbf{c})\,(\mathbf{b} \cdot \mathbf{d}). \tag{3.20}$$

Let f be a scalar field, $\mathbf{u}, \mathbf{v}$ be vector fields and $\mathbb{S}$ be a tensor field, all of them supposed to be smooth. Then, we have the following relations:

$$\begin{aligned}
\text{div }(f\mathbf{v}) &= f\,\text{div}\,\mathbf{v} + \mathbf{v} \cdot \boldsymbol{\nabla} f,\\
\text{div }(\mathbf{u} \otimes \mathbf{v}) &= \mathbf{u}\,\text{div}\,\mathbf{v} + (\boldsymbol{\nabla}\mathbf{u})\mathbf{v},\\
\text{div}(\mathbb{S}^\intercal \mathbf{u}) &= \mathbb{S} \cdot \boldsymbol{\nabla}\mathbf{u} + \mathbf{u} \cdot \text{div}\,\mathbb{S},\\
\text{div}(f\mathbb{S}) &= \mathbb{S}\boldsymbol{\nabla} f + f\,\text{div}\,\mathbb{S},\\
\text{div}(\mathbb{S}\mathbf{u}) &= \mathbf{u} \cdot \text{div}\,\mathbb{S}^\intercal + tr\ (\mathbb{S}\ \boldsymbol{\nabla}\mathbf{u}),\\
\text{div}(\mathbf{u} \times \mathbf{v}) &= (\text{curl }\ \mathbf{u}) \cdot \mathbf{v} - (\text{curl }\ \mathbf{v}) \cdot \mathbf{u}.
\end{aligned} \tag{3.21}$$

These relations may be verified either by the direct notation or bye the index notation. It turns out that it is not simple to express some of these relations using the direct notation (see, for example, Gurtin[Gurtin, 1982], pp. 30 and 31). However, it will become very simple if we express them in terms of the index notation, as we show in the sequel.

Finally, the *Laplacian* of $\mathbf{F}$ (where $\mathbf{F}$ is either a vector field or a scalar field $\mathbf{F} = \phi$ with continuous second-order partial derivatives) is denoted by $\triangle\mathbf{F}$ and defined by

$$\triangle\mathbf{F} = \text{div}(\boldsymbol{\nabla}\mathbf{F}). \tag{3.22}$$

In particular, if $\triangle\phi = 0$, we say that ϕ is *harmonic*.

3.1.4 *Gradient of a field in curvilinear coordinates*

If Ω is an open subset of $\mathbb{R}^n$ and $f : \Omega \to W$ is a differentiable tensor field, we can construct a function $\widehat{f}$ defined on an open subset of $\mathbb{R}^n$ and taking values on W by setting

$$f(\mathbf{x}) = \widehat{f}\left(x^1, \ldots, x^n\right),$$

where $x^1, \ldots, x^n$ are the curvilinear coordinates of $\mathbf{x}$. Since f is differentiable, it follows from (3.3) that

$$\begin{aligned}\nabla f(\mathbf{x})\ (\mathbf{v}) &= \frac{d}{dt}\widehat{f}(x^1 + tv^1, \ldots, x^n + tv^n)\bigg|_{t=0} \\ &= \frac{\partial \widehat{f}}{\partial x^i}\left(x^1, \ldots, x^n\right) v^i \\ &= \frac{\partial f}{\partial x^i}\left(\mathbf{x}\right) v^i. \end{aligned} \tag{3.23}$$

This formula will be specialized for the following cases:

a) If f is a differentiable scalar field $\phi : \Omega \to \mathbb{R}$, then (3.23) can be written as

$$\nabla\phi(\mathbf{x}) \cdot \mathbf{v} = \frac{\partial \phi(\mathbf{x})}{\partial x^i} v^i.$$

Since $\mathbf{v}$ in V is arbitrary, the substitution $\mathbf{v} = v^i \mathbf{g}_i$ yields

$$\nabla\phi(\mathbf{x}) \cdot \mathbf{g}_i = \frac{\partial \phi(\mathbf{x})}{\partial x^i}.$$

Hence,

$$\nabla\phi(\mathbf{x}) = \frac{\partial \phi(\mathbf{x})}{\partial x^i}\mathbf{g}^i, \tag{3.24}$$

which is the component form of the gradient of $\phi(\mathbf{x})$.

b) If f is a differentiable vector field $\mathbf{w} : \Omega \to W = V$, then it follows from (3.23) that

$$\nabla\mathbf{w}(\mathbf{x})\ (\mathbf{v}) = \frac{\partial \mathbf{w}(\mathbf{x})}{\partial x^i} v^i.$$

Again, due to the arbitrariness of $\mathbf{v}$ in V, we can write

$$\nabla\mathbf{w}(\mathbf{x})\ \mathbf{g}_i(\mathbf{x}) = \frac{\partial \mathbf{w}(\mathbf{x})}{\partial x^i}.$$

This expression is equivalent to

$$\nabla\mathbf{w}(\mathbf{x}) = \frac{\partial \mathbf{w}(\mathbf{x})}{\partial x^i} \otimes \mathbf{g}^i(\mathbf{x}), \tag{3.25}$$

by virtue of the canonical isomorphism $U \otimes V = L(V; U)$, where $\dfrac{\partial \mathbf{w}(\mathbf{x})}{\partial x^i} \in U$ and $\mathbf{g}^i(\mathbf{x}) \in V^*$.

c) If $f = \mathbb{T}$ denotes a differentiable tensor field of order n on Ω, then the same arguments which have produced (3.25) give

$$\nabla\mathbb{T}(\mathbf{x})\mathbf{g}_i(\mathbf{x}) = \frac{\partial\mathbb{T}(\mathbf{x})}{\partial x^i}$$

and

$$\nabla\mathbb{T}(\mathbf{x}) = \frac{\partial\mathbb{T}(\mathbf{x})}{\partial x^i} \otimes \mathbf{g}^i(\mathbf{x}). \tag{3.26}$$

3.1.5 *Christoffel symbols and covariant differentiation of vectors*

Consider a coordinate system $\{x^i\}$ on an open subset Ω of an Euclidean point space and let $\{\mathbf{g}_i(\mathbf{x})\}$ and $\{\mathbf{g}^i(\mathbf{x})\}$ be its natural bases, respectively. We introduce the coefficients

$$\Gamma_{ijk}(\mathbf{x}) = \Gamma_{jik}(\mathbf{x}) \tag{3.27}$$
$$\Gamma^k_{ij}(\mathbf{x}) = \Gamma^k_{ji}(\mathbf{x}) \tag{3.28}$$

so that

$$\begin{aligned} \frac{\partial\mathbf{g}_i(\mathbf{x})}{\partial x^j} &= \Gamma_{ijk}\mathbf{g}^k(\mathbf{x}), \\ \frac{\partial\mathbf{g}_i(\mathbf{x})}{\partial x^j} &= \Gamma^k_{ij}\mathbf{g}_k(\mathbf{x}). \end{aligned} \tag{3.29}$$

Dot-multiplying each of the equations in (3.29) by $\mathbf{g}_k$ and by $\mathbf{g}^k$, respectively, we have

$$\frac{\partial\mathbf{g}_i}{\partial x^j} \cdot \mathbf{g}_k = \Gamma_{ijk} \tag{3.30}$$

and

$$\Gamma^k_{ij} = \frac{\partial\mathbf{g}_i}{\partial x^j} \cdot \mathbf{g}^k = -\mathbf{g}_i \cdot \frac{\partial\mathbf{g}^k}{\partial x^j}. \tag{3.31}$$

Hence, from (3.31) and (3.28) we obtain

$$\frac{\partial\mathbf{g}^i}{\partial x^j} = -\Gamma^i_{jk}\mathbf{g}^k, \tag{3.32}$$

where we have left out the argument $\mathbf{x}$ in the notation. The coefficients Γ_{ijk} and Γ^i_{jk} are called *Christoffel symbols of first and second kinds*, respectively.

The symmetry of the Christoffel symbols in (3.27) and (3.28) can easily be verified, since

$$\frac{\partial \mathbf{g}_i}{\partial x^j} = \frac{\partial^2 \mathbf{x}}{\partial x^i \partial x^j} = \frac{\partial \mathbf{g}_j}{\partial x^i}$$

and, from (3.29),

$$\Gamma_{ij\ell}\, \mathbf{g}^\ell = \Gamma_{ji\ell}\, \mathbf{g}^\ell \qquad \text{and} \qquad \Gamma_{ij}^{\ell}\, \mathbf{g}_\ell = \Gamma_{ji}^{\ell}\, \mathbf{g}_\ell.$$

This gives the symmetry in question.

Suppose that $\mathbf{v}(\mathbf{x})$ is a vector field and

$$\mathbf{v}(\mathbf{x}) = v^i(\mathbf{x})\mathbf{g}_i(\mathbf{x}) = v_i(\mathbf{x})\mathbf{g}^i(\mathbf{x}) \tag{3.33}$$

its representation. Differentiating (3.33) and taking into account (3.29) together with (3.32) lead to

$$\begin{aligned}
\frac{\partial \mathbf{v}}{\partial x^j} &= \frac{\partial v^i}{\partial x^j}\mathbf{g}_i + v^i \frac{\partial \mathbf{g}_i}{\partial x^j} \\
&= \frac{\partial v^i}{\partial x^j}\mathbf{g}_i + v^i \Gamma_{ij}^k \mathbf{g}_k \\
&= \left(\frac{\partial v^i}{\partial x^j} + v^k \Gamma_{jk}^i \right) \mathbf{g}_i \\
&= v^i_{;j}\mathbf{g}_i
\end{aligned} \tag{3.34}$$

and

$$\begin{aligned}
\frac{\partial \mathbf{v}}{\partial x^j} &= \frac{\partial v_i}{\partial x^j}\mathbf{g}^i + v_i \frac{\partial \mathbf{g}^i}{\partial x^j} \\
&= \frac{\partial v_i}{\partial x^j}\mathbf{g}^i - v_i \Gamma_{jk}^i \mathbf{g}^k \\
&= \left(\frac{\partial v_i}{\partial x^j} - v_k \Gamma_{ij}^k \right) \mathbf{g}^i \\
&= v_{i;j}\mathbf{g}^i,
\end{aligned} \tag{3.35}$$

respectively. We remind the reader that the symbol ( ); denotes the operation called the covariant derivative.

The component form of the gradient $\mathbf{v}(\mathbf{x})$ can be obtained by substituting (3.34) and (3.35) into (3.25). Therefore,

$$\nabla \mathbf{v} = v^i_{;j}\mathbf{g}_i \otimes \mathbf{g}^j \tag{3.36}$$

and

$$\nabla \mathbf{v} = v_{i;j}\mathbf{g}^i \otimes \mathbf{g}^j \tag{3.37}$$

where

$$v^i_{;j} = \frac{\partial v^i}{\partial x^j} + v^k\,\Gamma^i_{jk} \qquad \text{and} \qquad v_{i;j} = \frac{\partial v^i}{\partial x^j} - v_k\,\Gamma^k_{ij}.$$

Since $\nabla\mathbf{v}(\mathbf{x})$ is a second-order tensor at $\mathbf{x}$ (see Section 3.1.1), $v^i_{;j}$ and $v_{i;j}$ are its mixed and covariant components, respectively. Thus, the covariant derivative of the components of $\mathbf{v}(\mathbf{x})$ increases its order by one.

3.1.6 *Metric tensors and Christoffel symbols*

Now, we can determine the Christoffel symbols in a more convenient way. Taking the partial derivatives of the covariant components of the metric tensor $g_{ij} = \mathbf{g}_i \cdot \mathbf{g}_j$ with respect to x^k, and using (3.29), we get

$$\begin{aligned}\frac{\partial g_{ij}}{\partial x^k} &= \frac{\partial \mathbf{g}_i}{\partial x^k} \cdot \mathbf{g}_j + \mathbf{g}_i \cdot \frac{\partial \mathbf{g}_j}{\partial x^k} \\ &= \Gamma^r_{ik}\,\mathbf{g}_r \cdot \mathbf{g}_j + \mathbf{g}_i \cdot \Gamma^r_{jk}\,\mathbf{g}_r \\ &= \Gamma^r_{ik}\,g_{rj} + \Gamma^r_{jk}\,g_{ir}.\end{aligned} \tag{3.38}$$

The cyclic permutations of the indices (i, j, k) of this last equation provide three new equations. Addition of two of them and subtraction of the remaining one lead to

$$\Gamma^k_{ij} = \frac{1}{2} g^{kr} \left(\frac{\partial g_{ir}}{\partial x^j} + \frac{\partial g_{jr}}{\partial x^i} - \frac{\partial g_{ij}}{\partial x^r} \right), \tag{3.39}$$

where g^{kr} denotes the contravariant components of the metric tensor.

Let us find the transformation rule for the Christoffel symbols. For this, consider the coordinate transformation

$$x^{i'} = x^{i'}\left(x^1, x^2, x^3\right)$$

and its inverse

$$x^i = x^i\left(x^{1'}, x^{2'}, x^{3'}\right).$$

We have already seen that, under these coordinates transformation, the vectors of the natural basis transform according to

$$\mathbf{g}_{i'} = \frac{\partial x^j}{\partial x^{i'}} \mathbf{g}_j, \qquad \mathbf{g}^{i'} = \frac{\partial x^{i'}}{\partial x^j} \mathbf{g}^j. \tag{3.40}$$

Taking into account these relations (3.40), the transformation formula for Christoffel symbols can be obtained as follows:

$$
\begin{aligned}
\Gamma^{k'}_{i'j'} &= \mathbf{g}^{k'} \cdot \frac{\partial \mathbf{g}_{i'}}{\partial x^{j'}} = \frac{\partial x^{k'}}{\partial x^{\ell}} \mathbf{g}^{\ell} \cdot \frac{\partial}{\partial x^{j'}} \left(\frac{\partial x^{s}}{\partial x^{i'}} \mathbf{g}_{s} \right) \\
&= \frac{\partial x^{k'}}{\partial x^{\ell}} \mathbf{g}^{\ell} \left(\frac{\partial^2 x^{s}}{\partial x^{i'} \partial x^{j'}} \mathbf{g}_{s} + \frac{\partial x^{s}}{\partial x^{i'}} \frac{\partial \mathbf{g}_{s}}{\partial x^{r}} \frac{\partial x^{r}}{\partial x^{j'}} \right) \\
&= \frac{\partial x^{k'}}{\partial x^{\ell}} \frac{\partial^2 x^{\ell}}{\partial x^{i'} \partial x^{j'}} + \frac{\partial x^{k'}}{\partial x^{\ell}} \frac{\partial x^{s}}{\partial x^{i'}} \frac{\partial x^{r}}{\partial x^{j'}} \mathbf{g}^{\ell} \cdot \frac{\partial \mathbf{g}_{s}}{\partial x^{r}},
\end{aligned}
$$

which gives

$$
\Gamma^{k'}_{i'j'} = \frac{\partial x^{k'}}{\partial x^{\ell}} \frac{\partial^2 x^{\ell}}{\partial x^{i'} \partial x^{j'}} + \frac{\partial x^{k'}}{\partial x^{\ell}} \frac{\partial x^{s}}{\partial x^{i'}} \frac{\partial x^{r}}{\partial x^{j'}} \Gamma^{\ell}_{rs}, \tag{3.41}
$$

where $\Gamma^{k'}_{i'j'}$ and Γ^{ℓ}_{rs} are the Christoffel symbols associated with the systems of coordinates $(x^{i'})$ and $(x^{j'})$, respectively. Since (3.41) does not obey the tensor transformation rule, it follows that the Christoffel symbols can not be taken as components of a third-order tensor.

We now return to the subject of the preceding subsection in order to extend formulas (3.36) and (3.37) to tensor fields. For this purpose, let us first solve the following example.

Example 3.3 Suppose that $\mathbb{T}$ is a second-order tensor represented by

$$
\mathbb{T} = T^{i}_{\cdot j} \mathbf{g}_{i} \otimes \mathbf{g}^{j}. \tag{3.42}
$$

Find the covariant derivative of $T^{i}_{\cdot j}$.

Solution: By differentiating (3.42) with respect to x^k and using (3.29) and (3.32), we have

$$
\begin{aligned}
\frac{\partial \mathbb{T}}{\partial x^{k}} &= T^{i}_{\cdot j,k} \mathbf{g}_{i} \otimes \mathbf{g}^{j} + T^{i}_{\cdot j} \frac{\partial \mathbf{g}_{i}}{\partial x^{k}} \otimes \mathbf{g}^{j} + T^{i}_{\cdot j} \mathbf{g}_{i} \otimes \frac{\partial \mathbf{g}^{j}}{\partial x^{k}} \\
&= T^{i}_{\cdot j,k} \mathbf{g}_{i} \otimes \mathbf{g}^{j} + T^{i}_{\cdot j} \Gamma^{r}_{ik} \mathbf{g}_{r} \otimes \mathbf{g}^{j} - T^{i}_{\cdot j} \mathbf{g}_{i} \otimes \Gamma^{j}_{\ell k} \mathbf{g}^{\ell} \\
&= \left(T^{i}_{\cdot j,k} + T^{s}_{\cdot j} \Gamma^{i}_{sk} - T^{i}_{\cdot \ell} \Gamma^{\ell}_{jk} \right) \mathbf{g}_{i} \otimes \mathbf{g}^{j} = T^{i}_{\cdot j;k} \mathbf{g}_{i} \otimes \mathbf{g}^{j}.
\end{aligned} \tag{3.43}
$$

The component form of *grad* $\mathbb{T}$ can be obtained by substituting (3.43) into (3.26). Therefore,

$$
\boldsymbol{\nabla} \mathbb{T} = T^{i}_{\cdot j;k} \mathbf{g}_{i} \otimes \mathbf{g}^{j} \otimes \mathbf{g}^{k}, \tag{3.44}
$$

where $\nabla\mathbb{T}$ is of order $(1,2)$ and

$$T^{i}_{\cdot j;k} = T^{i}_{\cdot j,k} + T^{s}_{\cdot j}\Gamma^{i}_{sk} - T^{i}_{\cdot \ell}\Gamma^{\ell}_{jk}, \tag{3.45}$$

is the covariant derivative of $T^{i}_{\cdot j}$.

Now, let $\mathbb{T}$ be a tensor field of order (p,q), namely,

$$\mathbb{T} = T^{i_1\cdots i_p}{}_{j_1\cdots j_q}\,\mathbf{g}_{i_1}\otimes\cdots\otimes\mathbf{g}_{i_p}\otimes\mathbf{g}^{j_1}\otimes\cdots\otimes\mathbf{g}^{j_q}. \tag{3.46}$$

Then, following a similar procedure as above, we obtain

$$\nabla\mathbb{T} = T^{i_1\cdots i_p}{}_{j_1\cdots j_q;k}\,\mathbf{g}_{i_1}\otimes\cdots\otimes\mathbf{g}_{i_p}\otimes\mathbf{g}^{j_1}\otimes\cdots\otimes\mathbf{g}^{j_q}\otimes\mathbf{g}^{k}, \tag{3.47}$$

where $\nabla\mathbb{T}$ is of order $(p,q+1)$ and $T^{i_1\cdots i_p}{}_{j_1\cdots j_q;k}$ is equal to

$$\begin{aligned} &T^{i_1\cdots i_p}{}_{j_1\cdots j_q,k} + T^{r i_2\cdots i_p}{}_{j_1\cdots j_q}\,\Gamma^{i_1}_{rk} + \cdots + T^{i_1\cdots i_{p-1} r}{}_{j_1\cdots j_q}\,\Gamma^{i_p}_{rk} \\ &\quad - T^{i_1\cdots i_p}{}_{r j_2\cdots j_q}\,\Gamma^{r}_{\cdot j_1,k} - \cdots - T^{i_1\cdots i_p}{}_{j_1\cdots j_{q-1} r}\,\Gamma^{r}_{\cdot j_q k} \end{aligned} \tag{3.48}$$

which is the covariant derivative of $T^{i_1\cdots i_p}{}_{j_1\cdots j_q}$.

If $\mathbb{G}=\mathbb{T}$ is the metric tensor, we know that its components are all constant in rectangular Cartesian coordinates. Thus, $\dfrac{\partial g_{ij}}{\partial x^k} = 0$. Consequently, the corresponding Christoffel symbols are zero, and the covariant derivatives of the metric tensor vanish in these coordinate systems. Therefore, the covariant derivatives of the metric tensor vanish in all coordinate systems, i.e.,

$$g_{ij;k} = g^{ij}_{;k} = g^{i}_{j;k} = 0, \tag{3.49}$$

a result known as Ricci's theorem.

Similarly, since the permutation symbols and the Krönecker deltas are constant in any rectangular Cartesian system of coordinates, we conclude that

$$\epsilon_{rst;i} = \epsilon^{rst}_{;i} = \delta^{r}_{s;i} = 0 \tag{3.50}$$

in every coordinate system.

A direct consequence of these results is the fact that the components of the metric tensor, the permutation symbols, and the Krönecker deltas commute with the operator of covariant differentiation.

3.1.7 *Other differential operators in curvilinear coordinates*

In order to find the component form of the divergence of a smooth vector field $\mathbf{v}$, we first make the substitution $\nabla\mathbf{v} = v^{i}_{;j}\mathbf{g}_i\otimes\mathbf{g}^j$ in (3.15) to get

$$\operatorname{div}\mathbf{v} = tr\,(\nabla\mathbf{v}) = v^{i}_{;i}. \tag{3.51}$$

Next, we rewrite

$$v^i_{;i} = \frac{\partial v^i}{\partial x^i} + v^k \Gamma^i_{ki} \tag{3.52}$$

using a simplified formula for the Christoffel symbols Γ^i_{ki}, which is obtained as follows. By Ricci's theorem, we have

$$g_{ij;k} = \frac{\partial g_{ij}}{\partial x^k} - \Gamma^r_{ki} g_{rj} - \Gamma^r_{kj} g_{ir} = 0.$$

Multiplication by g^{ij} gives

$$g^{ij} \frac{\partial g_{ij}}{\partial x^k} - \Gamma^i_{ki} - \Gamma^j_{kj} = 0,$$

that is,

$$\Gamma^i_{ki} = \frac{1}{2} g^{ij} \frac{\partial g_{ij}}{\partial x^k}.$$

Since

$$g = g_{ij} a^{ij} \qquad \text{and} \qquad g^{ij} = \frac{a^{ij}}{g},$$

where g denotes the determinant $|g_{ij}|$ and a^{ij} denotes the cofactor of the element g_{ij} in the determinant g, we have

$$\frac{\partial g}{\partial x^k} = a^{ij} \frac{\partial g_{ij}}{\partial x^k} = g g^{ij} \frac{\partial g_{ij}}{\partial x^k}.$$

Therefore,

$$\Gamma^i_{ki} = \frac{1}{2g} \frac{\partial g}{\partial x^k} = \frac{1}{\sqrt{g}} \frac{\partial (\sqrt{g})}{\partial x^k}. \tag{3.53}$$

Finally, inserting (3.53) in (3.52) gives

$$\operatorname{div} \mathbf{v} = v^i_{;i} = \frac{1}{\sqrt{g}} \frac{\partial}{\partial x^i} \left(\sqrt{g}\, v^i \right)$$

so that

$$\operatorname{div} \mathbf{v} = \frac{1}{\sqrt{g}} \frac{\partial}{\partial x^i} \left(\sqrt{g}\, g^{ik} v_k \right). \tag{3.54}$$

Note that, for rectangular Cartesian coordinates, the components of the metric tensor are all constant. Hence $\Gamma^i_{ki} = 0$ and formula (3.52) reduces to

$$\operatorname{div} \mathbf{v} = v_{i,i} = \frac{\partial v_i}{\partial \bar{x}^i} = \frac{\partial v_1}{\partial x} + \frac{\partial v_2}{\partial y} + \frac{\partial v_3}{\partial z}, \tag{3.55}$$

where we recall that we are representing the Cartesian coordinates either by (x^1, x^2, x^3) or (x, y, z).

Example 3.4 Show that

$$\operatorname{div} \mathbf{g}_i = \frac{1}{\sqrt{g}} \frac{\partial \sqrt{g}}{\partial x^i}.$$

Solution: First, note that, for the vector field $\mathbf{g}$, formula (3.25) reads

$$\nabla \mathbf{g}_i = \frac{\partial \mathbf{g}_i}{\partial x^k} \otimes \mathbf{g}^k,$$

where

$$\frac{\partial \mathbf{g}_i}{\partial x^k} = \Gamma^j_{ik} \mathbf{g}_j$$

[see formula (3.29)]. Thus, we have that $\nabla \mathbf{g}_i = \Gamma^j_{ik} \mathbf{g}_j \otimes \mathbf{g}^k$ and $\operatorname{div} \mathbf{g}_i = tr\ \nabla \mathbf{g}_i = \Gamma^k_{ik}$. Finally, from (3.53) we can write

$$\operatorname{div} \mathbf{g}_i = \Gamma^k_{ki} = \frac{1}{\sqrt{g}} \frac{\partial (\sqrt{g})}{\partial x^i}.$$

In order to determine the component form of the divergence of a second-order tensor field $\mathbb{S}$, consider $\mathbb{S} = S^{ij} \mathbf{g}_i \otimes \mathbf{g}_j$ and $\mathbf{a} = a_r \mathbf{g}^r$. Therefore,

$$\mathbb{S}^\mathsf{T} = S^{ij} \mathbf{g}_j \otimes \mathbf{g}_i \quad \text{and} \quad \mathbb{S}^\mathsf{T} \mathbf{a} = S^{ij} a_i \mathbf{g}_j. = B^j \mathbf{g}_j,$$

where $B^j = S^{ij} a_i$. Substituting

$$\nabla(\mathbb{S}^\mathsf{T} \mathbf{a}) = \nabla(B^j \mathbf{g}_j) = B^j_{;s} \mathbf{g}_j \otimes \mathbf{g}^s \tag{3.56}$$

in the last member of $(div\ \mathbb{S}) \cdot \mathbf{a} = div\ (\mathbb{S}^\mathsf{T} \mathbf{a}) = tr\ \nabla(\mathbb{S}^\mathsf{T} \mathbf{a})$, which was obtained by combining (3.16) with (3.15), we have

$$(div\ \mathbb{S}) \cdot a_i \mathbf{g}^i = tr\ (B^j_{;s} \mathbf{g}_j \otimes \mathbf{g}^s) = B^j_{;j} = a_i S^{ij}_{;j}. \tag{3.57}$$

Thus,

$$div\ \mathbb{S} = S^{ij}_{;j} \mathbf{g}_i. \tag{3.58}$$

More generally, the component form of the divergence of the tensor field of order (p, q)

$$\mathbb{T} = T^{i_1 \cdots i_k \cdots i_p}{}_{j_1 \cdots j_q}\, \mathbf{g}_{i_1} \otimes \cdots \otimes \mathbf{g}_{i_k} \otimes \cdots \otimes \mathbf{g}_{i_p} \otimes \mathbf{g}^{j_1} \otimes \cdots \otimes \mathbf{g}^{j_q} \tag{3.59}$$

with respect to the kth contravariant index is the tensor field of order $(p-1, q)$:

$$div\ \mathbb{T} = T^{i_1 \cdots k \cdots i_p}{}_{j_1 \cdots j_q;k}\ \mathbf{g}_{i_1} \otimes \cdots \otimes \mathbf{g}_{i_{k-1}} \otimes \mathbf{g}_{i_{k+1}} \otimes \mathbf{g}_{i_p} \otimes \mathbf{g}^{j_1} \otimes \cdots \otimes \mathbf{g}^{j_q}. \quad (3.60)$$

The component form of the Laplacian of a scalar field can be obtained by substituting $\mathbf{v} = \boldsymbol{\nabla}\phi$ and $v_k = \dfrac{\partial \phi}{\partial x^k}$ into (3.54), taking into account the definition (3.22). Thus,

$$\Delta\phi = div\ (\boldsymbol{\nabla}\phi) = \frac{1}{\sqrt{g}} \frac{\partial}{\partial x^i} \left(\sqrt{g}\ g^{ik} \frac{\partial \phi}{\partial x^k} \right). \quad (3.61)$$

A generalization to the Laplacian of a tensor field $\mathbb{T}$ of order (p, q), with $p \geq 0$ and $q \geq 0$, reads

$$\Delta\mathbb{T} = g^{rs}\ T^{i_1 \cdots i_p}{}_{j_1 \cdots j_q;rs}\ \mathbf{g}_{i_1} \otimes \cdots \otimes \mathbf{g}_{i_p} \otimes \mathbf{g}^{j_1} \otimes \cdots \otimes \mathbf{g}^{j_q}, \quad (3.62)$$

where $\mathbb{T}$ is given by (3.59). This expression is the component form of the Laplacian of $\mathbb{T}$, obtained from the definition

$$\Delta\mathbb{T} = div\ (\boldsymbol{\nabla}\mathbb{T}), \quad (3.63)$$

where $\Delta\mathbb{T}$ is a tensor field of the same order (p, q). For a vector field $\mathbf{v}$, the formula (3.62) reduces to

$$\Delta\mathbf{v} = g^{rs} v^i_{;rs} \mathbf{g}_i. \quad (3.64)$$

With the aid of the following two examples we will deduce the component form of the curl of a vector field starting from its coordinate-free definition (3.17). In the next example, we use the concept of double scalar product (:) of two tensor products of vectors defined by (3.20).

Example 3.5 Prove that

$$\mathbb{E} : \mathrm{grad}\,\mathbf{u} = -\ \mathbb{E} : (\mathrm{grad}\,\mathbf{u})^{\mathsf{T}}, \quad (3.65)$$

where $\mathbb{E}$ is the permutation tensor defined by (3.18).

Solution: Taking (3.18) and (3.25) into account and performing the operation of double scalar product of the left-hand side of (3.65), we have

$$\begin{aligned}
\mathbb{E} : \mathrm{grad}\,\mathbf{u} &= (\varepsilon_{rst}\mathbf{g}^r \otimes \mathbf{g}^s \otimes \mathbf{g}^t) : (\mathbf{u}_{,i} \otimes \mathbf{g}^i) \\
&= \varepsilon_{rst}\mathbf{g}^r (\mathbf{g}^s \cdot \mathbf{u}_{,i})(\mathbf{g}^t \cdot \mathbf{g}^i) \\
&= \varepsilon_{rts}\mathbf{g}^r (\mathbf{g}^t \cdot \mathbf{u}_{,i})(\mathbf{g}^s \cdot \mathbf{g}^i) \\
&= -\varepsilon_{rst}\mathbf{g}^r (\mathbf{g}^t \cdot \mathbf{u}_{,i})(\mathbf{g}^s \cdot \mathbf{g}^i) \\
&= -\mathbb{E} : (\mathbf{g}^i \otimes \mathbf{u}_{,i}) \\
&= -\mathbb{E} : (\mathrm{grad}\,\mathbf{u})^{\mathsf{T}}.
\end{aligned}$$

It should be noted that the substitution of (3.65) in (3.17) leads to

$$\operatorname{curl}\mathbf{u} = \frac{1}{2}\mathbb{E} : ((\operatorname{grad}\mathbf{u})^\mathsf{T} - \operatorname{grad}\mathbf{u}) = \mathbb{E} : (\operatorname{grad}\mathbf{u})^\mathsf{T}, \tag{3.66}$$

which will be used in the sequel in order to obtain the component form of $\operatorname{curl}\mathbf{u}$.

Example 3.6 Prove that the component form of $\operatorname{curl}\mathbf{u}$ is given by

$$\operatorname{curl}\mathbf{u} = \varepsilon^{rst}\mathbf{g}_r\mathbf{u}_{t;s}. \tag{3.67}$$

Solution: After substituting in the last member of (3.66) the permutation tensor written in the form

$$\mathbb{E} = \varepsilon^{rst}\mathbf{g}_r \otimes \mathbf{g}_s \otimes \mathbf{g}_t$$

and

$$(\operatorname{grad}\mathbf{u})^\mathsf{T} = \mathbf{g}^i \otimes \mathbf{u}_{,i} \tag{3.68}$$

we find

$$\begin{aligned}
\operatorname{curl}\mathbf{u} &= (\varepsilon^{rst}\mathbf{g}_r \otimes \mathbf{g}_s \otimes \mathbf{g}_t) : (\mathbf{g}^i \otimes \mathbf{u}_{,i}) \\
&= \varepsilon^{rst}\mathbf{g}_r(\mathbf{g}_s \cdot \mathbf{g}^i)(\mathbf{g}_t \cdot \mathbf{u}_{,i}) \\
&= \varepsilon^{rst}\mathbf{g}_r\mathbf{g}_s^i(\mathbf{g}_t \cdot \mathbf{g}^j u_{j;i}) \\
&= \varepsilon^{rst}\mathbf{g}_r(\mathbf{g}_s^i g_t^j u_{j;i}) \\
&= \varepsilon^{rst}\mathbf{g}_r u_{t;s},
\end{aligned}$$

which is the desired result.

In terms of generic coordinates $(x^1, x^2, x^3) \in \mathbb{R}^3$, we can write

$$\begin{aligned}
\operatorname{curl}\mathbf{u} &= \varepsilon^{ijk}\mathbf{g}_i u_{k;j} \\
&= \sum_{i,jk} \frac{\mathbf{g}_i}{\sqrt{g}}(u_{k;j} - u_{j;k})
\end{aligned} \tag{3.69}$$

after replacing r, s, and t by i, j, and k, respectively, in (3.67). The indices i, j, k are cyclic permutations of the numbers $1, 2, 3$. Also, we have $g = \det[g_{ij}]$. By virtue of the relations

$$u_{k;j} = u_{k,j} - u_m\Gamma^m_{kj} \quad \text{and} \quad u_{j;k} = u_{j,k} - u_m\Gamma^m_{jk}$$

we have $u_{k;j} - u_{j;k} = u_{k,j} - u_{j,k}$. Thus,

$$\operatorname{curl}\mathbf{u} = \sum_{i,jk} \frac{g_i}{\sqrt{g}}(u_{k,j} - u_{j,k}) = \varepsilon^{ijk} u_{k;j}\mathbf{g}_i \tag{3.70}$$

so that

$$\operatorname{curl}\mathbf{u} = \varepsilon^{ijk} u_{k;j}\mathbf{g}_i. \tag{3.71}$$

In rectangular Cartesian coordinates (x, y, z), since $g = \det[g_{ij}] = 1$ and the permutation symbols are identical with the components of the permutation tensor, this formula reads

$$\operatorname{curl}\mathbf{u} = \left(\frac{\partial u_3}{\partial y} - \frac{\partial u_2}{\partial z}\right)\mathbf{i} + \left(\frac{\partial u_1}{\partial z} - \frac{\partial u_3}{\partial x}\right)\mathbf{j} + \left(\frac{\partial u_2}{\partial x} - \frac{\partial u_1}{\partial y}\right)\mathbf{k}, \tag{3.72}$$

where u_i, $i = 1, 2, 3$ are the components of $\mathbf{u}$ in the directions of the rectangular axis, i.e., $\mathbf{u} = u_1\mathbf{i} + u_2\mathbf{j} + u_3\mathbf{k}$.

Now we are able to go back to relations (3.21) and express them in terms of the index notation. In fact, applying the product rules of differentiation, we have

$$\begin{aligned}
(fu^i)_{;i} &= f_{;i}u^i + fu^i_{;i}, \\
(u^i v^j)_{;j} &= u^i_{;j}v^j + u^i v^j_{;j}, \\
(S^{ji}u_j)_{;i} &= S^{ji}_{;i}u_j + S^{ji}u_{j;i}, \\
(fS^{ij})_{;j} &= f_{;j}S^{ij} + fS^{ij}_{;j}, \\
(S^{ij}u_j)_{;i} &= S^{ij}_{;i}u_j + S^{ij}u_{j;i}, \\
(\varepsilon^{ijk}u_j v_k)_{;i} &= \varepsilon^{ijk}u_{j;i}v_k + \varepsilon^{ijk}u_j v_{k;i},
\end{aligned} \tag{3.73}$$

where the terms inside the parentheses in the left-hand side of relations above are the tensor components of $(f\mathbf{u}), (\mathbf{u} \otimes \mathbf{v}), (\mathbb{S}^{\mathsf{T}}\mathbf{u}), (f\mathbb{S}), (\mathbb{S}\mathbf{u})$, and $(\mathbf{u} \times \mathbf{v})$, respectively.

Observe that if rectangular Cartesian coordinates are used, then there is no difference between covariant and contravariant indices. Furthermore, the components of the metric tensor are constants and, consequently, the Christoffel symbols are zero so that the covariant derivatives become partial derivatives.

The following two examples illustrate the use of rectangular Cartesian coordinates.

Example 3.7 Let $\mathbf{r}$ be the position vector of a point (x, y, z) in a tridimensional Euclidean space, with $r = |\mathbf{r}| = (x^2 + y^2 + z^2)^{1/2}$. Determine:

(a) ∇r.
(b) ∇r^n, where n is a real number.
(c) $\nabla(1/r)$.

Solution: (a) The gradient $\nabla r = \nabla(x^2+y^2+z^2)^{1/2}$ in a rectangular Cartesian system of coordinates is given by

$$\begin{aligned}\nabla r &= \frac{\partial}{\partial x}(x^2+y^2+z^2)^{\frac{1}{2}}\mathbf{i} + \frac{\partial}{\partial y}(x^2+y^2+z^2)^{\frac{1}{2}}\mathbf{j} + \frac{\partial}{\partial z}(x^2+y^2+z^2)^{\frac{1}{2}}\mathbf{k} \\ &= \frac{1}{2}\frac{2x}{(x^2+y^2+z^2)^{\frac{1}{2}}}\mathbf{i} + \frac{1}{2}\frac{2y}{(x^2+y^2+z^2)^{\frac{1}{2}}}\mathbf{j} + \frac{1}{2}\frac{2z}{(x^2+y^2+z^2)^{\frac{1}{2}}}\mathbf{k} \\ &= \frac{x}{r}\mathbf{i} + \frac{y}{r}\mathbf{j} + \frac{z}{r}\mathbf{k}\end{aligned}$$

so that we have

$$\nabla r = \frac{\mathbf{r}}{r} = \mathbf{e}_r, \tag{3.74}$$

which is the unit vector in the direction of $\mathbf{r}$.
(b) Similarly for the gradient of r^n, we have

$$\nabla r^n = \frac{\partial r^n}{\partial x}\mathbf{i} + \frac{\partial r^n}{\partial x}\mathbf{j} + \frac{\partial r^n}{\partial x}\mathbf{k} = \frac{\partial r^n}{\partial r}\frac{\partial r}{\partial x}\mathbf{i} + \frac{\partial r^n}{\partial r}\frac{\partial r}{\partial y}\mathbf{j} + \frac{\partial r^n}{\partial r}\frac{\partial r}{\partial z}\mathbf{k} = nr^{n-1}\nabla r.$$

Using (3.74), we obtain

$$\nabla r^n = nr^{n-2}\mathbf{r} = nr^{n-1}\mathbf{e}_r. \tag{3.75}$$

(c) The substitution $n=-1$ in (3.75) yields

$$\nabla\left(\frac{1}{r}\right) = -\frac{\mathbf{r}}{r^3} = -\frac{1}{r^2}\mathbf{e}_r. \tag{3.76}$$

Example 3.8 Let $\mathbf{u}$ and $\mathbf{v}$ be two vector fields of class C^1. Using rectangular Cartesian coordinates, show that

$$\operatorname{div}(\mathbf{u}\times\mathbf{v}) = \mathbf{v}\cdot\operatorname{curl}\mathbf{u} - \mathbf{u}\cdot\operatorname{curl}\mathbf{v}. \tag{3.77}$$

Solution: Suppose

$$\mathbf{u} = u_1\mathbf{i} + u_2\mathbf{j} + u_3\mathbf{k} \quad \text{and} \quad \mathbf{v} = v_1\mathbf{i} + v_2\mathbf{j} + v_3\mathbf{k}. \tag{3.78}$$

Then, we have

$$\mathbf{u} \times \mathbf{v} = \det \begin{bmatrix} \mathbf{i} & \mathbf{j} & \mathbf{k} \\ u_1 & u_2 & u_3 \\ v_1 & v_2 & v_3 \end{bmatrix} = (u_2 v_3 - u_3 v_2)\mathbf{i} + (u_3 v_1 - u_1 v_3)\mathbf{j} + (u_1 v_2 - u_2 v_1)\mathbf{k}. \tag{3.79}$$

So, the divergence of (3.79) is

$$\begin{aligned} \operatorname{div}(\mathbf{u} \times \mathbf{v}) &= \frac{\partial}{\partial x}(u_2 v_3 - u_3 v_2) + \frac{\partial}{\partial y}(u_3 v_1 - u_1 v_3) + \frac{\partial}{\partial z}(u_1 v_2 - u_2 v_1) \\ &= v_1 \left(\frac{\partial u_3}{\partial y} - \frac{\partial u_2}{\partial z} \right) + v_2 \left(\frac{\partial u_1}{\partial z} - \frac{\partial u_3}{\partial x} \right) + v_3 \left(\frac{\partial u_2}{\partial x} - \frac{\partial u_1}{\partial y} \right) \\ &\quad - u_1 \left(\frac{\partial v_3}{\partial y} - \frac{\partial v_2}{\partial z} \right) - u_2 \left(\frac{\partial v_1}{\partial z} - \frac{\partial v_3}{\partial x} \right) u_3 \left(\frac{\partial v_2}{\partial x} - \frac{\partial v_1}{\partial y} \right). \end{aligned} \tag{3.80}$$

Taking into account (3.72) and (3.78), we obtain (3.77).

3.1.8 *Riemann–Christoffel tensor*

Since the covariant derivatives of tensors are again tensors, we can find their covariant derivatives to obtain new tensors, called the second covariant derivatives of the original tensors. So, let

$$v_{i;j} = v_{i,j} - v_r \Gamma^r_{ij}$$

be the covariant derivative of the vector component v_i and let

$$v_{i;jk} = (v_{i,j} - v_r \Gamma^r_{\ ,j})_{,k} - (v_{s,j} - v_r \Gamma^r_{ij})\Gamma^s_{ik} - (v_{i,s} - v_r \Gamma^r_{is})\Gamma^s_{kj} \tag{3.81}$$

be its second covariant derivative. After some calculations and simplifications, we obtain the following expression for the difference

$$v_{i;jk} - v_{i;kj} = v_r R^r_{ijk}, \tag{3.82}$$

where

$$R^r_{ijk} = \Gamma^r_{ik,j} - \Gamma^r_{ij,k} + \Gamma^r_{sj}\Gamma^s_{ik} - \Gamma^r_{sk}\Gamma^s_{ij}.$$

It can be shown that R^r_{ijk} is a fourth-order tensor, called the *Riemann–Christoffel tensor*. In a Cartesian coordinate system, the Christoffel symbols are equal to zero, that is, $\Gamma^k_{ij} = 0$, so that $R^r_{ijk} = 0$. Since this is a tensor equation, it is also zero in any coordinate systems of an Euclidean space.

Therefore, the equality $R^{r}_{.ijk} = 0$ proves that the order of two covariant differentiations is interchangeable and, furthermore, it is both a necessary and sufficient condition for a space to be Euclidean.

Example 3.9 Using the index notation, prove that
(a) $\operatorname{div}(\nabla \mathbf{v})^{\mathsf{T}} = \nabla(\operatorname{div} \mathbf{v})$.
(b) $\operatorname{curl} \ \operatorname{grad} \phi = 0$.

Solution: (a) Since the left-hand side of relation (a) can be expressed in terms of the index notation as $(v^j_{;k})_{;j}$, we have

$$(v^j_{;k})_{;j} = v^j_{;kj} = v^j_{;jk} = (v^j_{;j})_{;k}, \tag{3.83}$$

where the interchangeability of the order of the covariant differentiation was used.
(b) We have seen (formula (3.24)) that, for any differentiable scalar field ϕ,

$$\operatorname{grad} \phi = \nabla \phi = \phi_{,i} \mathbf{g}^i.$$

Since both common and covariant differentiations are identical for scalar fields, we can write

$$\mathbf{v} = \operatorname{grad} \phi = \phi_{,i} \mathbf{g}^i = \phi_{;i} \mathbf{g}^i = v_i \mathbf{g}^i,$$

where $\phi_{;i} = v_i$ are the vector components. Now, taking the curl of *grad* ϕ and using (3.69), we get

$$\operatorname{curl} \ \operatorname{grad} \phi = \varepsilon^{rst} v_{t;s} \mathbf{g}_r = \varepsilon^{rst} \phi_{;ts} \mathbf{g}_r, \tag{3.84}$$

which is equal to zero in virtue of the symmetry of $\phi_{;ts}$ with respect to the indices t and s.

3.2 Integral theorems for scalar and vector fields

The second part of the fundamental theorem of calculus for line integrals establishes that if φ is a scalar field defined on an open connected set $\Omega \subset \mathbb{R}^n$ such that the gradient $\nabla \varphi$ exists and is continuous and $A, B \in \mathbb{R}^n$ are the extreme points of a piecewise smooth curve C in Ω, then

$$\int_C \nabla \varphi \cdot d\mathbf{r} = \varphi(B) - \varphi(A). \tag{3.85}$$

The three integral theorems we will see in this section, known as Green's theorem in the plane and Stokes' theorem and Gauss' theorem for vector fields on $\mathbb{R}^3$, can be interpreted as generalizations of the fundamental theorem of calculus. Essentially, what we have in all these cases are statements establishing that the value of the integral of some differential operator applied to a field F over a set Ω is equal to the integral of F over the boundary of Ω.

Green's theorem on the plane:

Let Ω be the following region

$$\Omega = \left\{(x, y) \in \mathbb{R}^2;\ a \leq x \leq b \text{ and } \varphi_1(x) \leq y \leq \varphi_2(x)\right\},$$

where φ_1 and φ_2 are continuous functions defined on $[a, b]$ with $\varphi_1(x) \leq \varphi_2(x)$, for all $x \in [a, b]$. Let $\partial\Omega$ denote the boundary of Ω. Suppose the functions given by $u_1(x, y)$ and $u_2(x, y)$ and their derivatives $\dfrac{\partial u_1}{\partial y}, \dfrac{\partial u_2}{\partial x}$ are defined and continuous on $\Omega \cup \partial\Omega$. Then

$$\int_\Omega \left(\frac{\partial u_2}{\partial x} - \frac{\partial u_1}{\partial y}\right) dx dy = \oint_{\partial\Omega} u_1(x, y) dx + u_2(x, y) dy$$

where the line integral is taken around $\partial\Omega$ in the counterclockwise direction.

The divergence theorem, also known as Gauss' theorem, essentially states a conservation law by saying that the total amount of 'expansion' or "stretching" of a vector field $\mathbf{u}$ within a volume V is equal to the flux of $\mathbf{u}$ out of the surface S which encloses this volume.

Gauss' divergence theorem:

Let V be a volume in the 3-dimensional space bounded by an orientable closed surface S and let $\mathbf{n}$ be the unit exterior vector normal to S. Suppose the vector field $\mathbf{u} = u_1(x, y, z)\mathbf{i} + u_2(x, y, z)\mathbf{j} + u_3(x, y, z)\mathbf{k}$ is continuous and has continuous first partial derivatives on some open set containing $V \cup S$. Then

$$\int_V \operatorname{div} \mathbf{u}\ dx dy dz = \int_S \mathbf{u} \cdot \mathbf{n}\ dS.$$

Expressing $\mathbf{n} = \cos\alpha\ \mathbf{i} + \cos\beta\ \mathbf{j} + \cos\gamma\ \mathbf{k}$ in terms of the angles $\alpha = \arccos(\mathbf{n} \cdot \mathbf{i})$, $\beta = \arccos(\mathbf{n} \cdot \mathbf{j})$, and $\gamma = \arccos(\mathbf{n} \cdot \mathbf{k})$ that the normal $\mathbf{n}$ forms with each of the coordinate axes, we can write the integral formula in the following expanded form:

$$\int_V \left(\frac{\partial u_1}{\partial x} + \frac{\partial u_2}{\partial y} + \frac{\partial u_3}{\partial z}\right) dx dy dz = \int_S (u_1 \cos\alpha + u_2 \cos\beta + u_3 \cos\gamma)\ dS.$$

The Stokes' theorem expresses the surface integral of the curl of a vector field in terms of the integral of the vector field over the boundary of the surface of integration.

Stokes' theorem:

Let C be a closed curve which forms the boundary of a surface S. Suppose the vector field $u = u_1(x,y,z)i + u_2(x,y,z)j + u_3(x,y,z)k$ is defined and having first partial derivatives continuous on some open set containing $S \cup C$. *Then we have, in terms of the right-handed Cartesian coordinates,*

$$\int_S \left[\left(\frac{\partial u_3}{\partial y} - \frac{\partial u_2}{\partial z}\right)\mathbf{i} + \left(\frac{\partial u_1}{\partial z} - \frac{\partial u_3}{\partial x}\right)\mathbf{j} + \left(\frac{\partial u_2}{\partial x} - \frac{\partial u_1}{\partial y}\right)\mathbf{k}\right] \cdot \mathbf{n}\, dS$$
$$= \oint_C u_1(x,y,z)dx + u_2(x,y,z)dy + u_3(x,y,z)dz,$$

where $\mathbf{n}$ is the unit exterior vector normal to S and the curve C is traversed in the right-hand orientation.

This formula can be expressed in a compact form as

$$\int_S (\operatorname{curl}\mathbf{u}) \cdot \mathbf{n}\, dS = \oint_C \mathbf{u} \cdot d\mathbf{r}.$$

We have also the tensor form of the divergence theorem in curvilinear coordinates:

$$\int_V u^i_{;i}\, dV = \int_S u^i n_i\, dS.$$

To state the analogy with our introductory comment, notice that, with respect to formula (3.85), the set in question is the curve C, the differential operator is the gradient of the scalar field φ and, since the boundary ∂C is given by the extreme points of the curve, we can say that $\varphi(B) - \varphi(A)$ plays the role of the integral of φ over the boundary. In the three integral theorems stated above, the sets are either surfaces or volumes and the differential operators are the curl and the divergent.

3.3 Exercises

(1) Consider $S = \{\mathbf{v}_1, \mathbf{v}_2, \mathbf{v}_3\}$, where

$$\mathbf{v}_1 = \begin{Bmatrix} 1 \\ 1 \\ -1 \end{Bmatrix}, \quad \mathbf{v}_2 = \begin{Bmatrix} 1 \\ 0 \\ -1 \end{Bmatrix}, \quad \mathbf{v}_3 = \begin{Bmatrix} 0 \\ -1 \\ 0 \end{Bmatrix}.$$

Find a basis for *spanS*. *Hint*: Notice that $\{\mathbf{v}_1, \mathbf{v}_2\}$ is linearly independent.

(2) Find the dimension of

$$span\left\{\begin{Bmatrix} -2 \\ -1 \\ 1 \\ 0 \end{Bmatrix}, \begin{Bmatrix} -1 \\ 0 \\ 1 \\ 1 \end{Bmatrix}, \begin{Bmatrix} 2 \\ 2 \\ 0 \\ 2 \end{Bmatrix}, \begin{Bmatrix} 0 \\ 1 \\ 1 \\ 2 \end{Bmatrix}, \begin{Bmatrix} 1 \\ 1 \\ 0 \\ 1 \end{Bmatrix}\right\},$$

Answer: 2.

(3) Show that the set of all vectors of the form

$$\begin{Bmatrix} a \\ b \\ 0 \end{Bmatrix}, \quad a, b \in \mathbb{R},$$

with the usual addition and multiplication by a scalar, is a vector space over $\mathbb{R}$. *Hint*: Show that any vector in the set lies on the span of

$$\left\{\begin{Bmatrix} 1 \\ 0 \\ 0 \end{Bmatrix}, \begin{Bmatrix} 0 \\ 1 \\ 0 \end{Bmatrix}\right\}.$$

(4) Let $\mathbf{M}$ be the vector space of all 2×2 matrices with real entries. Show that

$$\mathbf{B} = \left\{\begin{bmatrix} 1 & 0 \\ 0 & 0 \end{bmatrix}, \begin{bmatrix} 0 & 1 \\ 0 & 0 \end{bmatrix}, \begin{bmatrix} 0 & 0 \\ 1 & 0 \end{bmatrix}, \begin{bmatrix} 0 & 0 \\ 0 & 1 \end{bmatrix}\right\}$$

is a basis for $\mathbf{M}$.

(5) (a) Given the basis $\mathbf{B} = \{\mathbf{g}_1, \mathbf{g}_2, \mathbf{g}_3\}$, where

$$\mathbf{g}_1 = \begin{Bmatrix} 1 \\ 1 \\ 1 \end{Bmatrix}, \quad \mathbf{g}_2 = \begin{Bmatrix} 2 \\ 3 \\ 2 \end{Bmatrix}, \quad \mathbf{g}_3 = \begin{Bmatrix} 3 \\ 3 \\ 4 \end{Bmatrix},$$

find its reciprocal basis.
(b) If $\mathbf{v} = 3\mathbf{g}_1 + \mathbf{g}_2 + 2\mathbf{g}_3$, find the covariant components of $\mathbf{v}$.
(c) Evaluate $g_{ik} = \mathbf{g}_i \cdot \mathbf{g}_k$.
Answer:

$$\mathbf{g}^1 = \begin{Bmatrix} 6 \\ -1 \\ 3 \end{Bmatrix}, \mathbf{g}^2 = \begin{Bmatrix} -1 \\ 1 \\ 0 \end{Bmatrix}, \mathbf{g}^3 = \begin{Bmatrix} -1 \\ 0 \\ 1 \end{Bmatrix}.$$

(6) Given

$$\mathbf{v} = \begin{Bmatrix} 3 \\ 0 \\ 4 \end{Bmatrix}, \quad \mathbf{w} = \begin{Bmatrix} -1 \\ 1 \\ 1 \end{Bmatrix},$$

find
(a) $\mathbf{v} \cdot \mathbf{w}$,
(b) $|\mathbf{v}|$ and $|\mathbf{w}|$,
(c) $|\mathbf{v} + \mathbf{w}|$,
(d) $|\mathbf{v}| + |\mathbf{w}|$.
(**Answers:** (a) 1, (b) 5 and $\sqrt{3}$, (c) $\sqrt{30}$, and (d) $5 + \sqrt{3}$).

(7) Let $\{\mathbf{g}_1 = \mathbf{i},\ \mathbf{g}_2 = \mathbf{j}\}$ and

$$\left\{ \overline{\mathbf{g}}_1 = \begin{Bmatrix} \cos\alpha \\ -\sin\alpha \end{Bmatrix}, \quad \overline{\mathbf{g}}_2 = \begin{Bmatrix} \sin\alpha \\ \cos\alpha \end{Bmatrix} \right\}$$

be two bases in $\mathbb{R}^2$. Show that the angle between $\mathbf{i}$ and $\overline{\mathbf{g}}_1$ is α and the angle between $\mathbf{j}$ and $\overline{\mathbf{g}}_2$ is also α.

(8) Show that the following set of vectors forms an orthogonal basis for $\mathbf{V} = \mathbb{R}^3$:

$$\left\{ \overline{\mathbf{g}}_1 = \begin{Bmatrix} 1 \\ 0 \\ 1 \end{Bmatrix}, \quad \overline{\mathbf{g}}_2 = \begin{Bmatrix} -1 \\ 0 \\ 1 \end{Bmatrix}, \quad \overline{\mathbf{g}}_3 = \begin{Bmatrix} 0 \\ 312 \\ 0 \end{Bmatrix} \right\}.$$

(9) Suppose that the transformation

$$\overline{x} = x \cdot x$$
$$\overline{y} = y \cdot y$$

connects the coordinate systems $(x^i) = (x, y)$ and $(\overline{x}^i) = (x, y)$.
(a) Compute the Jacobian matrix of the transformation as well as its inverse matrix.
(b) Calculate the vectors $\overline{\mathbf{g}}_1$ and $\overline{\mathbf{g}}_2$.
Answers:

$$(a) \quad J = \begin{bmatrix} 2x & 0 \\ 0 & 2y \end{bmatrix} \quad \text{and} \quad J^{-1} = \frac{1}{2} \begin{bmatrix} \frac{1}{x} & 0 \\ 0 & \frac{1}{y} \end{bmatrix}.$$

$$(b) \quad \overline{\mathbf{g}}_1 = \frac{1}{2x}\mathbf{i}, \quad \overline{\mathbf{g}}_2 = \frac{1}{2y}\mathbf{j}.$$

(10) The spherical coordinates (x^i) are connected to rectangular coordinates $(\overline{x}^i)$ via

$$\begin{aligned}\overline{x}^1 &= x^1 \sin x^2 \cos x^3,\\ \overline{x}^2 &= x^1 \sin x^2 \sin x^3,\\ \overline{x}^3 &= x^1 \cos x^2.\end{aligned}$$

Find:
(a) the natural basis of the spherical coordinates,
(b) the Jacobian matrix

$$J = \left[\frac{\partial \overline{x}}{\partial x}\right]_{3\times 3}, \quad \text{and}$$

(c) the metric coefficients in matrix form, that is, $G = J^T J$.
Answers: (a)

$$\begin{aligned}\mathbf{g}_1 &= \sin x^2 \cos x^3 \mathbf{i} + \sin x^2 \sin x^3 \mathbf{j} + \cos x^2 \mathbf{k},\\ \mathbf{g}_2 &= x^1(\cos x^2 \cos x^3 \mathbf{i} + \cos x^2 \sin x^3 \mathbf{j} - \sin x^2 \mathbf{k}),\\ \mathbf{g}_3 &= x^1 \sin x^2(-\sin x^3 \mathbf{i} + \cos x^3 \mathbf{j}).\end{aligned}$$

(b)

$$J = \begin{bmatrix} \sin x^2 \cos x^3 & x^1 \cos x^2 \cos x^3 & -x^1 \sin x^2 \sin x^3 \\ \sin x^2 \sin x^3 & x^1 \cos x^2 \sin x^3 & -x^1 \sin x^2 \cos x^3 \\ \cos x^2 & -x^1 \sin x^2 & 0 \end{bmatrix}.$$

(c)

$$G = J^T J = \begin{bmatrix} 1 & 0 & 0 \\ 0 & (x^1)^2 & 0 \\ 0 & 0 & (x^1 \sin x^2)^2 \end{bmatrix},$$

that is, $g_{11} = 1$, $g_{22} = (x^1)^2$, and $g_{33} = (x^1 \sin x^2)^2$.

(11) The parabolic coordinates (u, v, ϕ) are connected to rectangular orthonormal Cartesian coordinates $(\bar{x}^i) = (x, y, z)$ by

$$\begin{aligned}x &= uv \cos\phi,\\ y &= uv \sin\phi,\\ z &= \frac{1}{2}(v^2 - u^2).\end{aligned}$$

Find the metric coefficients g_{ij}.
Answer: $g_{ij} = 0$, if $i \neq j$, $g_{11} = g_{22} = (u^2 + v^2)^2$, and $g_{33} = (u^2 v^2)^2$.

(12) Calculate the volume element dV corresponding to the parabolic coordinates (u, v, ϕ) given by

$$x = uv\cos\phi,$$
$$y = uv\sin\phi,$$
$$z = \frac{1}{2}(v^2 - u^2).$$

Answer: $dV = (uv\sqrt{u^2 + v^2})dudvd\phi$.

(13) The toroidal coordinates (r, θ, ϕ) are connected to rectangular orthonormal Cartesian coordinates $(\bar{x}^i) = (x, y, z)$ by

$$x = (a - r\cos\theta)\cos\phi,$$
$$y = (a - r\cos\theta)\sin\phi,$$
$$z = r\sin\phi,$$

where $a > 0$ is a given parameter. Find the metric coefficients g_{ij}.
Answer: $g_{ij} = 0$, if $i \neq j$, and $g_{11} = 1$, $g_{22} = r^2$, and $g_{33} = (a - r\cos\theta)^2$.

(14) Let $\{\mathbf{i}, \mathbf{j}, \mathbf{k}\}$ be a right-handed orthonormal basis. Show that, for

$$\mathbf{A} = A_1\mathbf{i} + A_2\mathbf{j} + A_3\mathbf{k},$$
$$\mathbf{B} = B_1\mathbf{i} + B_2\mathbf{j} + B_3\mathbf{k},$$

we have that $\mathbf{A} \times \mathbf{B} = C_1\mathbf{i} + C_2\mathbf{j} + C_3\mathbf{k}$, where

$$\begin{Bmatrix} C_1 \\ C_2 \\ C_3 \end{Bmatrix} = \begin{bmatrix} 0 & -A_3 & A_2 \\ A_3 & 0 & -A_1 \\ -A_2 & A_1 & 0 \end{bmatrix} \begin{Bmatrix} B_1 \\ B_2 \\ B_3 \end{Bmatrix}.$$

(15) Show that

$$(\mathbf{a} \cdot \mathbf{b} \times \mathbf{c})(\mathbf{d} \cdot \mathbf{e} \times \mathbf{f}) = \begin{vmatrix} \mathbf{a} \cdot \mathbf{d} & \mathbf{a} \cdot \mathbf{e} & \mathbf{a} \cdot \mathbf{f} \\ \mathbf{b} \cdot \mathbf{d} & \mathbf{b} \cdot \mathbf{e} & \mathbf{b} \cdot \mathbf{f} \\ \mathbf{c} \cdot \mathbf{d} & \mathbf{c} \cdot \mathbf{e} & \mathbf{c} \cdot \mathbf{f} \end{vmatrix}.$$

(16) Compute the volume element dV corresponding to the prolate spheroidal coordinates (u, v, ϕ) given by

$$x = a\sinh y \sin v \cos\phi,$$
$$y = a\sinh u \sin v \sin\phi,$$
$$z = a\cosh u \cos v.$$

Answer: $dV = \left(a^2\sqrt{\sinh^2 u + \sin^2 v}\right) \sinh u \sin v du dv d\phi$.

(17) Let $\mathbf{T} \in V \otimes W$, $\mathbf{T} \neq \mathbf{0}$. Show that in general there exist several representations of the form

$$\mathbf{T} = \mathbf{v}_i \otimes \mathbf{w}_i, \quad \mathbf{v}_i \in V, \ \ \mathbf{w}_i \in W$$

with $\{\mathbf{v}_1, \ldots, \mathbf{v}_n\}$ and $\{\mathbf{w}_1, \ldots, \mathbf{w}_n\}$ each linearly independent.

(18) Let $\mathbf{a} \in V$ and $\mathbf{b} \in W$ be such that $\mathbf{b} \otimes \mathbf{a} \neq \mathbf{0}$. Prove that

$$\mathbf{b} \otimes \mathbf{a} = \mathbf{w} \otimes \mathbf{v} \ \text{ if and only if } \ \mathbf{v} = \lambda\mathbf{a} \text{ and } \mathbf{w} = \lambda\mathbf{b} \text{ for some } \lambda \neq 0.$$

(19) Show that a self-contraction over two contravariant or two covariant indices of a tensor does not give another tensor. For example, T_{abc} is not a tensor.

(20) Show that $L(V, L(V, V))$ is isomorphic to $T_2^1(V)$.

(21) Prove that $g_{ij}\mathbf{A}^i\mathbf{B}^j$ is invariant, i.e., $g_{\bar{i}\bar{j}}\mathbf{A}^{\bar{i}}\mathbf{B}^{\bar{j}} = g_{ij}\mathbf{A}^i\mathbf{B}^j$.

(22) Associated with the vectors $\mathbf{a} = a_i\mathbf{e}^i$ $\mathbf{b} = b_j\mathbf{e}^j$, and $\mathbf{c} = c_k\mathbf{e}^k$ it is possible to define the following trilinear functional $\mathbf{T}$ by putting

$$\mathbf{T}(\mathbf{a}, \mathbf{b}, \mathbf{c}) = a_i b_j c_k T_k^{ij}.$$

Show that this functional defines a tensor.

(23) Associated with the standard basis $\{\mathbf{e}_1, \mathbf{e}_2, \mathbf{e}_3\}$ of $\mathbb{R}^3$,

$$\mathbf{e}_1 = \begin{Bmatrix} 1 \\ 0 \\ 0 \end{Bmatrix}, \quad \mathbf{e}_2 = \begin{Bmatrix} 0 \\ 1 \\ 0 \end{Bmatrix}, \quad \mathbf{e}_3 = \begin{Bmatrix} 0 \\ 0 \\ 1 \end{Bmatrix},$$

there exists a set of scalars defined by $T^i_{jk} = T(\mathbf{e}^i, \mathbf{e}_j, \mathbf{e}_k)$, where T is a given function and $\mathbf{e}^i$ is the reciprocal defined by $\mathbf{e}^i \cdot \mathbf{e}_j = \delta^i_j$. Suppose another basis is defined as

$$\mathbf{e}_{\bar{1}} = \mathbf{e}_1 - 2\mathbf{e}_2,$$
$$\mathbf{e}_{\bar{2}} = 2\mathbf{e}_1 + \mathbf{e}_2,$$
$$\mathbf{e}_{\bar{3}} = \mathbf{e}_1 + \mathbf{e}_2.$$

Find $T^{\bar{i}}_{\bar{j}\bar{k}} = T(\mathbf{e}^{\bar{i}}, \mathbf{e}_{\bar{j}}, \mathbf{e}_{\bar{k}})$ in terms of T^i_{jk}.

(24) Show that $T^i_j\mathbf{v}_i\mathbf{w}^j$ is invariant.

(25) Show that if $a_j = T_{ij}\mathbf{v}^i$ are the components of a covariant vector $\mathbf{a}$ for all contravariant vectors $\mathbf{v}$, then T_{ij} is a covariant tensor of order 2.

(26) If v^i are the components of a contravariant vector, when can you say that a^i defined by $a^i = v^i + \alpha v_i$, $\alpha \in \mathbb{R}$, are components of a vector $\mathbf{a}$?

Chapter 4

Physical and Anholonomic Components of Tensors

The anholonomic tensor calculus is well founded componentwise on a change of bases and not of coordinates (since they do not exist). On the other hand, the holonomic tensor calculus, which is founded on the idea of coordinate transformation, may as well be founded on the notion of transformation of bases. Thus, the crucial notion here is the transformation of bases and not of coordinates. Therefore, the change of bases allows for the unification of both calculus on one context only, which gathers together all known operations of the tensor calculus, such as covariance and contravariance, invariance of the tensor expressions, raising and lowering of indexes, component contraction, and covariant derivatives.

In general, the covariant and contravariant components of a vector differ from the components used in physics. To see this, consider for instance the components of the velocity $\mathbf{v}$ in a system of spherical coordinates, that is,

$$v^r = \frac{\partial r}{\partial t}, \quad v^\theta = \frac{\partial \theta}{\partial t}, \quad v^\phi = \frac{\partial \phi}{\partial t}. \tag{4.1}$$

Considering that $\dim v^r = LT^{-1}$ and $\dim v^\theta = T^{-1}$ and $\dim v^\phi = T^{-1}$, these vector components are not dimensionally homogeneous, that is, they do not always have the dimension for velocity LT^{-1}. On the other hand, considering that the dimension of tensors are defined in terms of the dimensions of their components in an orthonormal Cartesian system of coordinates, the same sort of discrepancies are bound to be present. To remove this drawback, the tensor must have another sort of components that should be dimensionally homogeneous in any given system of coordinates whilst maintaining the tensor properties. In [Altman and Oliveira, 1977], some aspects of the theory of physical and anholonomic components of tensors are presented by associating these components to the spaces of linear transformations and tensor products, respectively. More specifically, two different methods associated with the theory developed by the authors in the above mentioned references can be used to determine the physical components of tensors, both of them yielding the same result. They are outlined next.

(1) The *physical component method*, which is based upon the "Truesdell's hypothesis."In this case, each tensor is seen as a linear transformation between two vector spaces properly chosen. The concept of physical components depends fundamentally on this linear transformation and on the unit bases of the mentioned vector spaces as well as their dual bases.
(2) The *method of invariance of the tensor representation*, which does not need the "Truesdell's hypothesis." The point here is that such components are obtained by means of the invariance of the tensorial representation along with the decomposition of the tensor on both anholonomic and natural bases. Examples of this procedure will be given further on in this chapter.

4.1 Physical and anholonomic components of vectors

Consider a body occupying, at a time $t \geq 0$, a region C of a three-dimensional Euclidean point space $\mathbf{E}$. Any point $\mathrm{P} \in C$ is called a *material point* of the body and the region $\mathcal{C}$ is referred to as the *reference configuration*. Strictly speaking, a *configuration* of the body at time t is a one-to-one mapping which takes material points in C to points in $\mathbf{E}$. We choose the reference configuration C to be the image of the initial configuration, say, at $t = 0$. Based on this choice, we say that the body occupying a region in $\mathbf{E}$ at any time $t > 0$ is in the current or *deformed configuration*, and we denote this region by c. The position of a material point in the reference configuration c is given by the curvilinear coordinates X^J, $(J = 1, 2, 3)$, while the position of the same point in the deformed configuration $\mathcal{C}$ is specified by new curvilinear coordinates x^j, $(j = 1, 2, 3)$. Note that the distinction is specified in terms of taking capital letters for the reference configuration and small letters for the deformed configuration.

The direct and reciprocal bases associated with each point X in C, respectively denoted $\mathbf{G} = \{\mathbf{G}_J(\mathbf{X})\}$ and $\mathbf{G}^* = \{\mathbf{G}^J(\mathbf{X})\}$, together with their corresponding metrics $(G_{IJ}(\mathbf{X}), G^{IJ}(\mathbf{X}))$, are defined by

$$\mathbf{G}_J = \mathbf{R}_{,J}, \quad G_{IJ} = \mathbf{G}_I \cdot \mathbf{G}_J = \mathbf{G}_J \cdot \mathbf{G}_I = G_{JI} \quad (I, J = 1, 2, 3) \tag{4.2}$$

$$\mathbf{G}^I = G_J^{IJ}\mathbf{G}, \quad G^{IJ} = \mathbf{G}^I \cdot \mathbf{G}^J = \mathbf{G}^J \cdot \mathbf{G}^I = G^{JI} = \frac{G}{D}. \tag{4.3}$$

In these expressions, G denotes the cofactors of G_{IJ} in the expansions of the determinants $D = \det(G_{IJ})$ and $\mathbf{R}$ is the position vector

$$\mathbf{R} = \mathbf{R}(X^1, X^2, X^3),$$

which we assume to be a single-valued function of the curvilinear coordinates and to have continuous derivatives up to any desired order at all points of the region where they are defined. The indices after a comma indicate differentiation with respect to X^J. These vector fields form what we call *natural bases* of the reference configuration and their corresponding metrics $(G_{IJ}(\mathbf{X}), G^{IJ}(\mathbf{X}))$ are called *natural metrics* of the reference configuration.

The counterparts of the bases $\mathbf{G}$ and $\mathbf{G}^*$ associated with each point x in c, denoted

$$\mathbf{g} = \{\mathbf{g}_j(\mathbf{x})\} \quad \text{and} \quad \mathbf{g}^* = \{\mathbf{g}^{\cdot j}(\mathbf{x})\},$$

and their corresponding metrics ($g_{IJ}(\mathbf{x})$ and $g^{IJ}(\mathbf{x})$) are defined in a similar way by

$$\mathbf{g}_j = \mathbf{r}_{,j}, \quad g_{ij} = \mathbf{g}_i \cdot \mathbf{g}_j = \mathbf{g}_j \cdot \mathbf{g}_i = g_{ji} \tag{4.4}$$

$$\mathbf{g}^{\,i} = g^{ij}_{j}\mathbf{g}, \quad g^{ij} = \mathbf{g}^{\,i} \cdot \mathbf{g}^{\,j} = \mathbf{g}^{\,j} \cdot \mathbf{g}^{\,i} = g^{ji} = \frac{g}{d} \tag{4.5}$$

for $i, j = 1, 2, 3$. Again, g denotes the cofactors g_{ij} in the expansions of the determinants $d = \det(g_{ij})$ and the position vector $\mathbf{r}(x^1, x^2, x^3)$ is assumed to be a single-valued function of the curvilinear coordinates and have continuous derivatives up to any desired order at all points of the region where they are defined. The indices after a comma indicate differentiation with respect to x^j. The nomenclature natural bases and natural metrics apply to the deformed configurations similarly. The numbers $g_{ij} = \mathbf{g}_i \cdot \mathbf{g}_j$, which we call **metric coefficients**, are important quantities that characterize the geometric properties of the coordinate system. It should be noted that, whilst the magnitudes and directions of the vectors of the Cartesian basis are fixed, the basis vectors $\mathbf{g}_i$'s in general vary from point to point in the space. Thus, in the sequel we will assume that each $\mathbf{g}_i$ is a vector field defined on some open set Ω contained in $\mathfrak{E}$, the Euclidean space of points.

Definition 4.1 The **reciprocal basis** of $\{\mathbf{g}_i\}$ at each point $\mathbf{x} \in \Omega$ is the basis $\{\mathbf{g}^i\}$ such that

$$\mathbf{g}^{\,i} \cdot \mathbf{g}_i = \delta^i_j, \quad i, j = 1, 2, \cdots, n.$$

These equations define completely the vectors $\mathbf{g}^i$, which are called the **contravariant basis vectors**. The vectors $\mathbf{g}_i$ are called **covariant basis vectors**.

With respect to the bases $\{\mathbf{g}_i\}$ and $\{\mathbf{g}^i\}$, a vector $\mathbf{v}$ can be represented as

$$\mathbf{v} = v^i\mathbf{g}_i = v_i\mathbf{g}^i,$$

where v^i and v_i are the *contravariant and covariant components*, respectively, of the vector $\mathbf{v}$. These components can be determined from the following relations

$$\mathbf{v} \cdot \mathbf{g}^i = v^j \mathbf{g}_j \cdot \mathbf{g}^i = v^j \delta^i_j = v^i,$$
$$\mathbf{v} \cdot \mathbf{g}_i = v_j \mathbf{g}^j \cdot \mathbf{g}_i = v_j \delta^j_i = v_i.$$

Thus,

$$\mathbf{v} = (\mathbf{v} \cdot \mathbf{g}^i)\mathbf{g}_i = (\mathbf{v} \cdot \mathbf{g}_i)\mathbf{g}^i. \tag{4.6}$$

The dot product of $\mathbf{v}$ with $\mathbf{g_i}$ and $\mathbf{g^i}$ leads to, respectively,

$$v_k = g_{ik} v^i \quad \text{and} \quad v^k = g^{ik} v_i,$$

where

$$g_{ik} = \mathbf{g}_i \cdot \mathbf{g}_k = g_{ki} \quad \text{and} \quad g^{ik} = \mathbf{g}^i \cdot \mathbf{g}^k = g^{ki}.$$

With respect to the bases $\{\mathbf{g}^k\}$ and $\{\mathbf{g}_k\}$, the reciprocity yields

$$\mathbf{g}_i = g_{ik}\, \mathbf{g}^k \quad \text{and} \quad \mathbf{g}^i = g^{ik}\, \mathbf{g}_k$$

and their dot product with $\mathbf{g}^j$ and $\mathbf{g}_j$ gives, respectively,

$$g^{ik} g_{kj} = \delta^i_j \quad \text{and} \quad g_{ik} g^{kj} = \delta^j_i$$

in a procedure that is sometimes known as the process of the *raising and lowering of indices.*

Let us now associate to a point $\mathbf{x} \in \Omega$ two coordinate systems, namely $x^{i'}$ and x^i, related to each other by the following transformations:

$$x^{i'} = x^{i'}(x^1, x^2, \cdots, x^n), \quad i' = 1, 2, \cdots, n,$$
$$x^i = x^i(x^{1'}, x^{2'}, \cdots, x^{n'}), \quad i = 1, 2, \cdots, n,$$

which are assumed to be continuously differentiable with nonzero Jacobians. The following statement holds for an arbitrary displacement vector $d\mathbf{x}$ with respect to the natural bases $\{\mathbf{g}_i\}$ and $\{\mathbf{g}_{i'}\}$ of the coordinate systems x^i and $x^{i'}$, respectively,

$$d\mathbf{x} = \mathbf{g}_i\, dx^i = \mathbf{g}_{i'}\, dx^{i'}, \quad i, i' = 1, 2, \cdots, n. \tag{4.7}$$

Also, we define the **arc element** ds by the formula

$$(ds)^2 = d\mathbf{x} \cdot d\mathbf{x} = g_{ij}\, dx^i\, dx^j. \tag{4.8}$$

Multiplying (4.7) by $\mathbf{g}^j$ and $\mathbf{g}^{j'}$, respectively, we obtain

$$dx^j = \mathbf{g}^j \cdot \mathbf{g}_{i'}\, dx^{i'}, \quad j = 1, 2, \cdots, n, \tag{4.9}$$

$$dx^{j'} = \mathbf{g}^{j'} \cdot \mathbf{g}_i dx^i, \quad j' = 1, 2, \cdots, n. \tag{4.10}$$

On the other hand, since

$$dx^j = \frac{\partial x^j}{\partial x^{i'}}\, dx^{i'} \quad \text{and} \quad dx^{j'} = \frac{\partial x^{j'}}{\partial x^i}\, dx^i,$$

for $j, j' = 1, 2, \cdots, n$, the comparison of these two sets of relations with (4.9) and (4.10) yields

$$\mathbf{g}^i \cdot \mathbf{g}_{i'} = \frac{\partial x^j}{\partial x^{i'}},$$

$$\mathbf{g}^{j'} \cdot \mathbf{g}_i = \frac{\partial x^{j'}}{\partial x^i}.$$

Consequently, we can write

$$\mathbf{g}_{i'} = \frac{\partial x^j}{\partial x^{i'}} \mathbf{g}_i, \quad \mathbf{g}_i = \frac{\partial x^{j'}}{\partial x^i} \mathbf{g}_{j'},$$

$$\mathbf{g}^{j'} = \frac{\partial x^{j'}}{\partial x^i} \mathbf{g}^i, \quad \mathbf{g}^j = \frac{\partial x^j}{\partial x^{j'}} \mathbf{g}^{i'}.$$

Furthermore, the transformation laws for the covariant and contravariant components can be written as

$$v_{i'} = \frac{\partial x^j}{\partial x^{i'}}\, v_j, \quad v_i = \frac{\partial x j'}{\partial x^i}\, v_{j'},$$

$$v^{j'} = \frac{\partial x^{j'}}{\partial x^i}\, v^i, \quad v^j = \frac{\partial x^j}{\partial x^{i'}}\, v^{i'}.$$

As an illustration, the metric coefficients of the spherical and cylindrical coordinate systems are presented here.

a) *Spherical coordinates*:

$$ds^2 = dr^2 + (r d\theta)^2 + (r \sin\theta d\varphi)^2,$$

$$g_{rr} = g^{rr} = 1, \quad g_{\theta\theta} = r^2, \quad g^{\theta\theta} = \frac{1}{r^2},$$

$$g_{\varphi\varphi} = (r \sin\theta)^2, \quad g^{\varphi\varphi} = \frac{1}{(r \sin\theta)^2},$$

and all the other components are zero.

b) *Cylindrical coordinates*:

$$ds^2 = dr^2 + (r d\varphi)^2 + dz^2,$$
$$g_{rr} = g^{rr} = 1, \quad g_{zz} = g^{zz} = 1,$$
$$g_{\varphi\varphi} = r^2, \quad g^{\varphi\varphi} = \frac{1}{r^2}.$$

and all the other components are zero.

Example 4.1
Suppose a coordinate transformation τ is given by

$$\bar{x} = x - y\cot\alpha, \quad \bar{y} = y\csc\alpha,$$

where $0 < \alpha < \pi$. (a) Find the Jacobian of the inverse transformation τ^{-1}. (b) Find the vectors $\mathbf{g}_1$ and $\mathbf{g}_2$ of the natural basis of the system $(\bar{x}, \bar{y})$. (c) Calculate the metric coefficients g_{ij}. (d) Assuming that (v^i) are the contravariant components of a vector in $\mathbb{R}^3$, find their correspondent covariant components v_i in the Euclidean metric $G = [g_{ij}]$.

Solution:
(a) The inverse transformation is easily seen to be

$$x = \bar{x} + \bar{y}\cos\alpha, \quad y = \bar{y}\sin\alpha.$$

Then, its Jacobian is given by

$$\frac{\partial(x,y)}{\partial(\bar{x},\bar{y})} = \det\begin{bmatrix} 1 & \cos\alpha \\ 0 & \sin\alpha \end{bmatrix} = \sin\alpha.$$

(b) We have that $\mathbf{g}_1$ and $\mathbf{g}_2$ are given by

$$\begin{Bmatrix} \mathbf{g}_1 \\ \mathbf{g}_2 \end{Bmatrix} = \begin{bmatrix} 1 & 0 \\ \cos\alpha & \sin\alpha \end{bmatrix} \begin{Bmatrix} \mathbf{i} \\ \mathbf{j} \end{Bmatrix}$$

so that

$$\mathbf{g}_1 = \mathbf{i}, \quad \mathbf{g}_2 = \mathbf{i}\cos\alpha + \mathbf{j}\sin\alpha.$$

(c) Concerning the metric coefficients, we have

$$g_{11} = \mathbf{g}_1 \cdot \mathbf{g}_1 = 1,$$
$$g_{12} = \mathbf{g}_1 \cdot \mathbf{g}_2 = \cos\alpha,$$
$$g_{22} = \mathbf{g}_2 \cdot \mathbf{g}_2 = 1.$$

Note that g_{12} is equal to the cosine of the angle between the vectors $\mathbf{g}_1$ and $\mathbf{g}_2$. The metric coefficients g_{ij} can be incorporated in the following matrix G, called Euclidean metric:

$$G = \begin{bmatrix} g_{11} & g_{12} \\ g_{21} & g_{22} \end{bmatrix}.$$

(d) The Euclidean metric is

$$G = \begin{bmatrix} 1 & \cos\alpha \\ \cos\alpha & 1 \end{bmatrix}.$$

Since $v_i = g_{ij}v^j$, we have

$$\begin{Bmatrix} v_1 \\ v_2 \end{Bmatrix} = \begin{bmatrix} 1 & \cos\alpha \\ \cos\alpha & 1 \end{bmatrix} \begin{Bmatrix} v^1 \\ v^2 \end{Bmatrix}$$

so that

$$v_1 = v^1 + v^2\cos\alpha, \quad v_2 = v^1\cos\alpha + v^2.$$

Notice that, for $\alpha = \pi/2$, the covariant and contravariant components coincide.

The following relations are satisfied by the vectors of the bases and the metrics associated with them:

$$\mathbf{G}^I \cdot \mathbf{G}_J = \delta^I_J \text{ and } G^{IK}G_{KJ} = \delta^I_J, \tag{4.11}$$

$$\mathbf{g}^i \cdot \mathbf{g}_j = \delta^i_j \text{ and } g^{ik}g_{kj} = \delta^i_j, \tag{4.12}$$

where δ^I_J and δ^i_j are the Kronecker delta symbols.

In the expansion of a vector field $\mathbf{V}(X)$ evaluated on a typical particle in C in the reference configuration,

$$\mathbf{V}(X) = V^I(X)\mathbf{G}_I(X) = V_I(X)\mathbf{G}^I(X), \tag{4.13}$$

we say that $V^I(X)$ and $V_I(X)$ are the *contravariant and covariant components of* $\mathbf{V}(X)$ with respect to the natural bases. If the vector field $\mathbf{V}$ is smooth on the region C, in the sense that it is continuously differentiable there, then the component form of the gradient of $\mathbf{V}$, denoted $\boldsymbol{\nabla}\mathbf{V}$, is given by

$$\boldsymbol{\nabla}\mathbf{V} = V^I_{;J}\mathbf{G}_I \otimes \mathbf{G}^J = V_{I;J}\mathbf{G}^I \otimes \mathbf{G}^J \tag{4.14}$$

with

$$V^I_{;\,J} = \frac{\partial V^I}{\partial X^J} + V^R \Gamma^I_{RJ}, \tag{4.15}$$

$$V_{I;J} = \frac{\partial V_I}{\partial X^J} - V_R \Gamma^R_{IJ}. \tag{4.16}$$

The notation $()_{;J}$ indicates the covariant partial derivative with respect to X^J, the symbol $\otimes$ denotes the tensor product, and Γ^K_{IJ} is the **Christoffel symbol** defined on $\mathcal{C}$ as

$$\Gamma^K_{IJ} = \mathbf{G}^K \cdot \frac{\partial \mathbf{G}_I}{\partial X^K}. \tag{4.17}$$

Formula (4.14) can also be rewritten as

$$\boldsymbol{\nabla}\mathbf{V} = \mathbf{V}_{,J} \otimes \mathbf{G}^J, \tag{4.18}$$

$$\mathbf{V}_{,J} = V^I_{;\,J}\mathbf{G}_I = V_{I\,;\,J}\mathbf{G}^I, \tag{4.19}$$

where $V^I_{;J}$ and $V_{I;J}$ denote the covariant partial derivatives with respect to X^J of a contravariant component V^I and of a covariant component V_I, respectively. The covariant derivative $T_{IK;J}$ of the component T_{IK} of a second-order tensor $\mathbb{T}$ with respect to X^J is

$$T_{IK;J} = \frac{\partial T_{IK}}{\partial X^J} - T_{MK}\Gamma^M_{IJ} - T_{IM}\Gamma^M_{KJ}. \tag{4.20}$$

Of course, similar formulas can be obtained for the deformed configuration by substituting the uppercase indices and letters which enter in the above expressions by lowercase ones. In the sequel we shall use alternatively each one of the undeformed (or reference) and deformed configurations and the exact situation will be specified by the proper notation.

We now proceed to formalize a precise characterization of natural (or holonomic) bases. Consider a tridimensional Cartesian system of coordinates with a fixed origin O. Since in a Cartesian system the vectors of the basis do not vary when the locus of the origin is changed, the correspondence between any point P and its position vector $\mathbf{r}$ relative to the origin allows us to identify P with the vector $\mathbf{r} = \overrightarrow{\mathrm{OP}}$. It is natural to take the basis made up of the vectors defined by

$$\mathbf{g}_i = \frac{\partial \mathbf{r}}{\partial x^i}, \quad i = 1, 2, 3, \tag{4.21}$$

where $\mathbf{r}$ is a single-valued function of the curvilinear coordinates (x^1, x^2, x^3). We usually denote such functional relations by writing $\mathbf{r} = \mathbf{r}(x^1, x^2, x^3)$. As we pass from a point P, with coordinates x^k, to a neighboring point, with coordinates $x^k + dx^k$, we have that

$$d\mathbf{r} = \mathbf{g}_k dx^k,$$

where the basis vectors $\mathbf{g}_k$ are tangent to their respective coordinate lines x^k. If the vectors $\mathbf{g}_k$ are assumed to be differentiable functions of the curvilinear coordinates x^k, the vectors of the basis $\mathbf{g} = (\mathbf{g}_1, \mathbf{g}_2, \mathbf{g}_3)$ must satisfy the integrability conditions

$$\frac{\partial \mathbf{g}_k}{\partial x^j} = \frac{\partial \mathbf{g}_j}{\partial x^k}, \quad i \neq j \ \ (i, j = 1, 2, 3).$$

Due to the fact that it is expressed by an integrability condition, this is therefore a holonomic constraint (i.e., concerned with the domain as a whole) which ensures the existence of a coordinate system (x_1, x_2, x_3) whose tangent vectors are $\mathbf{g}_i = \dfrac{\partial \mathbf{r}}{\partial x_i}$. With this motivation in mind, we define:

Definition 4.2 We say that an arbitrary ordered basis $\mathbf{g} = (\mathbf{g}_1, \mathbf{g}_2, \mathbf{g}_3)$ is a **holonomic** or **natural basis** iff

$$\frac{\partial \mathbf{g}_k}{\partial x^j} = \frac{\partial \mathbf{g}_j}{\partial x^k}, \quad i \neq j \ \ (1, 2, 3). \tag{4.22}$$

Later on we will see that if the vectors of a pair of reciprocal bases $\mathbf{g}$ and $\mathbf{g}^*$ are not all of unit length we may find some difficulties relating the covariant and contravariant components v_k and v^k,

$$\mathbf{v} = v^k \mathbf{g}_k \quad \text{and} \quad \mathbf{v} = v_k \mathbf{g}^k, \ \ k = 1, 2, \tag{4.23}$$

of the same vector $\mathbf{v}$. In order to overcome such problems, it is useful to construct unit bases (bases composed only of vectors of unit lengths) by setting

$$\mathbf{e}_a = \frac{1}{h_a}\mathbf{g}_a \quad \text{and} \quad \mathbf{f}^a = \frac{1}{h^a}\mathbf{g}^a \tag{4.24}$$

together with considering their reciprocal counterparts

$$\mathbf{e}^a = h_a \mathbf{g}^a \quad \text{and} \quad \mathbf{f}_a = h^a \mathbf{g}_a. \tag{4.25}$$

Here, the symbols h_a and h^a, known as *metric symbols*, stand for

$$h_a = \sqrt{g_{aa}} = \sqrt{\mathbf{g}_a \cdot \mathbf{g}_a}, \tag{4.26}$$

$$h^a = \sqrt{g^{aa}} = \sqrt{\mathbf{g}^a \cdot \mathbf{g}^a}, \tag{4.27}$$

respectively. We recall that, since the metric symbols contain double indices, they do not contribute to the summation convention. Thus, for instance, $h_a \mathbf{g}^a$ is not a sum of vectors but just a single vector $\mathbf{e}^a$.

Expanding the vector $\mathbf{v}$ with respect to each of these bases and taking into account the vector invariance, we write

$$\mathbf{v} = [v^a]_e \mathbf{e}_a, \tag{4.28}$$

$$\mathbf{v} = [v^a]_f \mathbf{f}_a, \tag{4.29}$$

$$\mathbf{v} = [v_a]_{e^*} \mathbf{e}^a, \tag{4.30}$$

$$\mathbf{v} = [v_a]_{f^*} \mathbf{f}^a, \tag{4.31}$$

$$\mathbf{v} = v^a \mathbf{g}_a, \tag{4.32}$$

$$\mathbf{v} = v_a \mathbf{g}^a, \tag{4.33}$$

with $a = 1, 2, 3$. On the other hand, recalling that the unit bases are no more than the normalization of the natural bases, we obtain from (4.28) and (4.32),

$$[v^a]_e \mathbf{e}_a = v^a \mathbf{g}_a,$$

$$[v^a]_e \frac{1}{h_a} \mathbf{g}_a = v^a \mathbf{g}_a.$$

From the uniqueness of representation that follows from the linear independence of the vectors of a basis, we get $[v^a]_e \frac{1}{h_a} = v^a$, that is, $[v^a]_e = h_a v^a$. Proceeding similarly for each case, we have the complete set of relations

$$[v^a]_e = h_a v^a \quad \text{and} \quad [v^a]_f = \frac{1}{h^a} v^a, \tag{4.34}$$

$$[v_a]_{f^*} = h^a v_a \quad \text{and} \quad [v_a]_{e^*} = \frac{1}{h_a} v_a. \tag{4.35}$$

Definition 4.3 The coefficients that appear multiplying the vector components in the above relations (4.34) and (4.35) are called **physical components** of the vector $\mathbf{v}$ relative to the bases $\mathbf{e} = (\mathbf{e}_1, \mathbf{e}_2, \mathbf{e}_3)$, $\mathbf{f} = (\mathbf{f}_1, \mathbf{f}_2, \mathbf{f}_3)$, $\mathbf{f}^* = (\mathbf{f}^1, \mathbf{f}^2, \mathbf{f}^3)$, and $\mathbf{e}^* = (\mathbf{e}^1, \mathbf{e}^2, \mathbf{e}^3)$, respectively. We use the bracket notation $[\cdot]_{\mathcal{B}}$ to denote physical components of a vector referred to a basis $\mathcal{B}$ of the considered vector space.

The constraint (4.22), that is,

$$\frac{\partial \mathbf{g}_k}{\partial x^j} = \frac{\partial \mathbf{g}_j}{\partial x^k}, \quad i \neq j \;\; (1, 2, 3),$$

is not a general rule because most basis systems of tangent vector spaces do not satisfy this condition. In order to go around this problem, we can introduce pseudodifferentials (or anholonomic differentials) dx^a in the following way. Put

$$dx^a = J_k^a dx^k, \tag{4.36}$$

where the matrix $J = [J_k^a]$ is such that $\det J \neq 0$ and dx^k is an exact differential. Each dx^a so defined is not necessarily an exact differential and from the invariance of the vector $d\mathbf{r}$ we have

$$d\mathbf{r} = dx^k \mathbf{g}_k = dx^a \mathbf{b}_a, \tag{4.37}$$

where $(\mathbf{g}_k)$ is a natural basis and $(\mathbf{b}_a)$ is an anholonomic covariant basis. Substituting relation (4.36) in (4.37) yields

$$d\mathbf{r} = dx^k \mathbf{g}_k = J_k^a \mathbf{b}_a dx^k \tag{4.38}$$

or

$$\mathbf{g}_k = J_k^a \mathbf{b}_a. \tag{4.39}$$

From (4.39) we can relate the anholonomic and natural bases by

$$\mathbf{b}_a = J_a^k \mathbf{g}_k, \tag{4.40}$$

and, similarly, we can write for the reciprocal bases,

$$\mathbf{b}^a = J_k^a \mathbf{g}^k. \tag{4.41}$$

Definition 4.4 A basis $\mathbf{e} = (\mathbf{e}_1, \mathbf{e}_2, \mathbf{e}_3)$ is called an **anholonomic basis** if and only if

$$\frac{\partial \mathbf{e}_a}{\partial x^b} \neq \frac{\partial \mathbf{e}_b}{\partial x^a} \quad \text{for some} \quad a, b \in \{1, 2, 3\}, \ a \neq b, \tag{4.42}$$

where

$$\frac{\partial}{\partial x^a} = J_a^k \frac{\partial}{\partial x^k} \quad \text{and} \quad \frac{\partial}{\partial x^b} = J_b^k \frac{\partial}{\partial x^k}. \tag{4.43}$$

Moreover, a contravariant basis of vectors $\mathbf{f}^* = (\mathbf{f}^1, \mathbf{f}^2, \mathbf{f}^3)$ is anholonomic if and only if

$$\frac{\partial \mathbf{f}^a}{\partial x^b} \neq \frac{\partial \mathbf{f}^b}{\partial x^a} \quad \text{for some} \quad a, b \in \{1, 2, 3\}, \ a \neq b. \tag{4.44}$$

The vectors of anholonomic bases $\mathbf{e}$ and $\mathbf{f}^*$ can be represented in terms of the natural bases $\mathbf{g} = (\mathbf{g}_1, \mathbf{g}_2, \mathbf{g}_3)$ and $\mathbf{g}^* = (\mathbf{g}^1, \mathbf{g}^2, \mathbf{g}^3)$ as

$$\mathbf{e}_a = e_a^k \mathbf{g}_k \quad \text{and} \quad \mathbf{f}^b = f_k^b \mathbf{g}^k, \quad k, a, b = 1, 2, 3, \tag{4.45}$$

respectively. The coefficients

$$e_a^k = \mathbf{e}_a \cdot \mathbf{g}^k \quad \text{and} \quad f_k^b = \mathbf{f}^b \cdot \mathbf{g}_k \tag{4.46}$$

are the components of $\mathbf{e}_a$ and $\mathbf{f}^b$ relative to the bases $\mathbf{g}$ and $\mathbf{g}^*$, respectively.

Similarly, we can represent the anholonomic bases $\mathbf{e}^*$ and $\mathbf{f}$ as

$$\mathbf{e}^a = e_k^a \mathbf{g}^k \quad \text{and} \quad \mathbf{f}_b = f_b^k \mathbf{g}_k, \tag{4.47}$$

where, because $\mathbf{e}^a \cdot \mathbf{g}_k = e_k^a \mathbf{g}^k \cdot \mathbf{g}_k = e_k^a$, we have that

$$e_k^a = \mathbf{e}^a \cdot \mathbf{g}_k, \tag{4.48}$$

and

$$f_b^k = \mathbf{f}_b \cdot \mathbf{g}^k \tag{4.49}$$

together with formulas (4.45) define the so-called *factors of basis transformation.*

The factors of basis transformation e_a^k, f_k^a, e_k^a, and f_a^k, associated with the anholonomic bases $\mathbf{e} = (\mathbf{e}_a)$, $\mathbf{f}^* = (\mathbf{f}^a)$, $\mathbf{e}^* = (\mathbf{e}^a)$, and $\mathbf{f} = (\mathbf{f}_a)$, respectively, satisfy the following relations:

$$e_j^a e_a^i = \delta_j^i \quad \text{and} \quad f_j^a f_a^i = \delta_j^i, \tag{4.50}$$

$$e_j^a e_b^j = \delta_b^a \quad \text{and} \quad f_j^a f_b^j = \delta_b^a, \tag{4.51}$$

with $i, j, a, b = 1, 2, 3$. In order to verify, say, the first relation in (4.50), to get the desired result we substitute

$$\mathbf{e}^a = e_j^a \mathbf{g}^j \quad \text{and} \quad \mathbf{e}_b = e_b^i \mathbf{g}_i \tag{4.52}$$

into the condition of reciprocity of the bases $\mathbf{e}$ and $\mathbf{e}^*$, i.e.,

$$\mathbf{e}^a \cdot \mathbf{e}_b = \delta_b^a = \begin{cases} 0 \text{ if } a \neq b \\ 1 \text{ if } a = b \end{cases}, \quad a, b = 1, 2, 3. \tag{4.53}$$

All the other relations (4.50) and (4.51) are obtained in a similar manner.

We note that since $\mathbf{b}$ is a generic basis, the pair of formulas (4.40) and (4.41) can be transformed into the corresponding formulas referred to the bases $\mathbf{e}$ or $\mathbf{f}$ by means of a suitable change of basis and factors of basis transformations. Also, we observe that all the above formulas can be expressed similarly in terms of bases of the set $\{\mathbf{E}, \mathbf{E}^*, \mathbf{F}, \mathbf{F}^*\}$ in the reference configuration. The following example illustrates the concepts introduced so far.

Example 4.2 Consider the transformation

$$x = \sin\theta\,\cos\phi, \quad y = \sin\theta\,\sin\phi, \quad z = \cos\theta, \tag{4.54}$$

and assume that $\rho = 1$, $0 < \theta < \pi$, $0 \leq \phi < \pi$. Show that this transformation defines a spherical system of coordinates where $\mathbf{r} = x\mathbf{i} + y\mathbf{j} + z\mathbf{k}$ is the position vector of $P(x, y, z)$ in Cartesian coordinates.

Solution: Let us first show that the unit basis vectors, denoted $\mathbf{e}_\theta$, $\mathbf{e}_\phi$, in spherical coordinates and referred to the current configuration, are

$$\mathbf{e}_\theta = \cos\theta\cos\phi\ \mathbf{i} + \cos\theta\sin\theta\ \mathbf{j} - \sin\theta\ \mathbf{k}, \tag{4.55}$$

$$\mathbf{e}_\phi = -\sin\phi\ \mathbf{i} + \cos\phi\ \mathbf{j}. \tag{4.56}$$

Writing the expression of the position vector $\mathbf{r}$ in terms of spherical coordinates, we have

$$\mathbf{r} = x\mathbf{i} + y\mathbf{j} + z\mathbf{k} = \sin\theta\cos\phi\ \mathbf{i} + \sin\theta\sin\phi\ \mathbf{j} + \cos\theta\ \mathbf{k}. \tag{4.57}$$

Consequently, the expressions of the natural basis vectors are given by

$$\mathbf{g}_\theta = \frac{\partial \mathbf{r}}{\partial \theta} = \cos\theta\cos\phi\ \mathbf{i} + \cos\theta\sin\phi\ \mathbf{j} - \sin\theta\ \mathbf{k}, \tag{4.58}$$

$$\mathbf{g}_\phi = \frac{\partial \mathbf{r}}{\partial \phi} = -\sin\theta\sin\phi\ \mathbf{i} + \sin\theta\cos\phi\ \mathbf{j}, \tag{4.59}$$

where

$$\left|\frac{\partial \mathbf{r}}{\partial \theta}\right| = \sqrt{\mathbf{g}_\theta \cdot \mathbf{g}_\theta} = 1, \quad \text{and} \quad \left|\frac{\partial \mathbf{r}}{\partial \phi}\right| = \sqrt{\mathbf{g}_\phi \cdot \mathbf{g}_\phi} = \sin\theta. \tag{4.60}$$

At this point, it is easy to verify that the basis vectors $\mathbf{g}_\theta$ and $\mathbf{g}_\phi$ satisfy the constraint condition of the natural or holonomic basis. In fact,

$$\frac{\partial \mathbf{g}_\phi}{\partial \theta} = \frac{\partial \mathbf{g}_\theta}{\partial \phi} = \cos\theta\ [-\sin\phi\ \mathbf{i} + \cos\phi\ \mathbf{j}]. \tag{4.61}$$

The substitution of the formulas (4.58) and (4.59) into

$$\mathbf{e}_\theta = \frac{\dfrac{\partial \mathbf{r}}{\partial \theta}}{\left|\dfrac{\partial \mathbf{r}}{\partial \theta}\right|} \quad \text{and} \quad \mathbf{e}_\phi = \frac{\dfrac{\partial \mathbf{r}}{\partial \phi}}{\left|\dfrac{\partial \mathbf{r}}{\partial \phi}\right|}. \tag{4.62}$$

and taking into account that

$$\left|\frac{\partial \mathbf{r}}{\partial \theta}\right| = 1, \quad \left|\frac{\partial \mathbf{r}}{\partial \phi}\right| = \sin\theta,$$

which are in agreement with formulas (4.60), leads to the unit vectors (4.55) and (4.56). Denoting

$$\mathbf{e}_\theta = \mathbf{e}_1, \quad \mathbf{g}_\theta = \mathbf{g}_1, \quad \mathbf{e}_\phi = \mathbf{g}_2, \quad \mathbf{g}_\phi = \mathbf{g}_2, \tag{4.63}$$

and, taking into account that $\mathbf{e}_\theta = \mathbf{g}_\theta$ are unit vectors, we can write

$$\mathbf{e}_1 = e_1^1 \mathbf{g}_1 \text{ where } e_1^1 = 1, \text{ since } \mathbf{e}_1 = \mathbf{g}_1. \tag{4.64}$$

Similarly,

$$\mathbf{e}_2 = e_2^2 \mathbf{g}_2 \text{ where } e_2^2 = \frac{1}{\sin\theta}, \text{ since } \mathbf{e}_2 = \frac{1}{\sin\theta}\mathbf{g}_2. \tag{4.65}$$

Next, in order to verify whether the basis $(\mathbf{e}_1, \mathbf{e}_2)$ is anholonomic, let us calculate the terms of the condition

$$\frac{\partial \mathbf{e}_2}{\partial x^{(1)}} \neq \frac{\partial \mathbf{e}_1}{\partial x^{(2)}}, \tag{4.66}$$

where

$$\frac{\partial}{\partial x^{(1)}} = \frac{\partial}{\partial \theta} \text{ and } \frac{\partial}{\partial x^{(2)}} = \frac{1}{\sin\theta}\frac{\partial}{\partial \phi}. \tag{4.67}$$

Hence,

$$\frac{\partial \mathbf{e}_2}{\partial x^{(1)}} = \frac{\partial \mathbf{e}_2}{\partial \theta}, \tag{4.68}$$

$$\frac{\partial \mathbf{e}_1}{\partial x^{(2)}} = \frac{1}{\sin\theta}\frac{\partial \mathbf{e}_1}{\partial \phi} = \frac{\cos\theta}{\sin\theta}\mathbf{e}_2 = \cot\theta\, \mathbf{e}_2, \tag{4.69}$$

which shows that the basis $(\mathbf{e}_1, \mathbf{e}_2)$ is anholonomic.

Since the physical components are special cases of anholonomic components, they can also be written as

$$[v^a]_e = e_k^a v^k \text{ and } [v^a]_f = f_k^a v^k, \tag{4.70}$$

$$[v_a]_{f^*} = f_a^k v_k \text{ and } [v_a]_{e^*} = e_a^k v_k, \tag{4.71}$$

where the factors of basis transformation are obtained after comparing formulas (4.70)–(4.71) with (4.24)–(4.25). Thus (no sum on k),

$$e_k^a = \delta_k^a\, h_k \text{ and } f_k^a = \frac{\delta_k^a}{h^k}, \tag{4.72}$$

$$f_a^k = \delta_a^k\, h^k \text{ and } e_a^k = \frac{\delta_a^k}{h_k}. \tag{4.73}$$

The physical components of a vector defined by formulas (4.28)–(4.31) are interpreted in Figure 4.1. This figure shows that the components OC = $[v^a]_e = e^a_k v^k$ and OB = $[v_a]_{f^*} = f^k_a v_k$ are parallel projections of the vector $\mathbf{v}$ onto the directions $\mathbf{g}_1$ and $\mathbf{g}^1$, respectively. The second components OD = $[v_a]_{e^*} = e^k_a v_k$ and OA = $[v^a]_f = f^a_k v^k$ are orthogonal projections over the same directions, respectively. It should be noted that the components referred to the bases $\mathbf{e}$ and $\mathbf{f}^*$ add by the parallelogram rule to form the vector itself. Furthermore, from the figure we have

$$[v_a]_{e^*} = [v_a]_{f^*} \cos(\mathbf{g}^k, \mathbf{g}_k), \tag{4.74}$$

$$[v^a]_f = [v^a]_e \cos(\mathbf{g}^k, \mathbf{g}_k). \tag{4.75}$$

Since the factors $\cos(\mathbf{g}^k, \mathbf{g}_k)$ are physically dimensionless, this shows that the components $[v_a]_{e^*}$ and $[v^a]_f$ have the same physical dimensions as the components $[v_a]_{f^*}$ and $[v^a]_e$, respectively. Thus, the components defined by formulas (4.70)–(4.71) have the same physical dimensions.

As shown in Figure 4.1, the parallel projections are

$$\text{OC} = [v^1]_e = \sqrt{g_{11}}\, v^1,$$

$$\text{OB} = [v_1]_{f^*} = \sqrt{g^{11}}\, v_1,$$

and the orthogonal projections are

$$\text{OD} = [v_1]_{e^*} = \frac{v_1}{\sqrt{g_{11}}},$$

$$\text{OA} = [v^1]_f = \frac{v^1}{\sqrt{g^{11}}}.$$

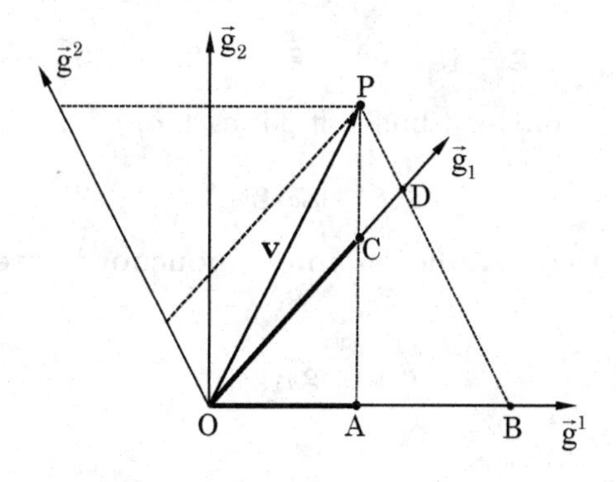

Fig. 4.1 Interpretation of the physical components.

We end this section by observing that we can introduce the natural and anholonomic bases by means of an alternative formalism. For this, assume that the position vector $\mathbf{x} = \mathbf{x}(x^1, x^2, \cdots, x^n)$ of a point P has continuous derivatives up to any desired order. We can write

$$\mathbf{dx} = dx^i \frac{\partial \mathbf{x}}{\partial x^i},$$

where the vectors $\frac{\partial \mathbf{x}}{\partial x^i}$ are tangent to the n coordinate lines intersecting at the point P. These vectors form the natural basis

$$\frac{\partial \mathbf{x}}{\partial x^i}, \quad i = 1, \cdots, n$$

at each point $\mathbf{x} \in \Omega$ (recall that $\mathbf{x}$ is the position vector of the point P).

Definition 4.5 Consider two arbitrary vector fields $\mathbf{v}_a$ and $\mathbf{v}_b$ represented in terms of an arbitrary basis $\mathbf{g} = \{\mathbf{g}_1, \ldots, \mathbf{g}_n\}$ as

$$\mathbf{v}_a = v^i_a \mathbf{g}_i,$$
$$\mathbf{v}_b = v^j_b \mathbf{g}_j.$$

The **commutator** of $\mathbf{v}_a$ and $\mathbf{v}_b$, denoted $[\mathbf{v}_a, \mathbf{v}_b]$, is the new vector field $\mathbf{c}_{ab} = C^k_{ab} \mathbf{g}_k$ whose kth component is given by

$$C^k_{ab} = [\mathbf{v}_a, \mathbf{v}_b]^k = \sum_{r=1}^{n} \left(v^r_a \partial_r v^k_b - v^r_b \partial_r v^k_a \right).$$

The expression $\partial_r v^k_b$ denotes the partial derivative of the function v^k_b with respect to the coordinate x^r.

Definition 4.6 We say that an arbitrary ordered basis

$$\mathbf{g} = \{\mathbf{g}_1, \mathbf{g}_2, ..., \mathbf{g}_a, ..., \mathbf{g}_b, \ldots, \mathbf{g}_n\}$$

is a **holonomic** or **natural basis** iff, for each $a, b = 1, \cdots, n$,

$$\mathbf{c}_{ab} = [\mathbf{g}_a, \mathbf{g}_b] = \mathbf{0}.$$

The basis is said to be a **anholonomic** or **noncoordinate basis** iff, for each $a, b = 1, \cdots, n$,

$$\mathbf{c}_{ab} = [\mathbf{g}_a, \mathbf{g}_b] \neq \mathbf{0}.$$

If we define

$$\mathbf{g}_a = \frac{\partial \mathbf{x}}{\partial x^a}, \quad \mathbf{g}_b = \frac{\partial \mathbf{x}}{\partial x^b}, \quad a, b = 1, \cdots, n,$$

then we have that

$$[\mathbf{g}_a, \mathbf{g}_b]^k = \mathbf{g}_a^k \partial_k \mathbf{g}_b - \mathbf{g}_b^k \partial_k \mathbf{g}_a = \partial_a \mathbf{g}_b - \partial_b \mathbf{g}_a = \mathbf{0},$$

since

$$\mathbf{g}_a^k = \frac{\partial x^k}{\partial x^a} = \delta_a^k,$$
$$\mathbf{g}_a^k \partial_k \mathbf{g}_b = \delta_a^k \partial_k \mathbf{g}_b = \partial_a \mathbf{g}_b,$$
$$\partial_a \mathbf{g}_b = \partial_b \mathbf{g}_a = \frac{\partial^2 \mathbf{x}}{\partial x^a \partial x^b}.$$

Thus, in this particular case, the basis $\mathbf{g} = \{\mathbf{g}_1, \mathbf{g}_2, ..., \mathbf{g}_a, ..., \mathbf{g}_\mathbf{b} \dots, \mathbf{g}_n\}$ is holonomic or natural and $[\mathbf{g}_a, \mathbf{g}_b] = \mathbf{0}$. On the other hand, the basis of polar unit vectors

$$\mathbf{e}_1 = \frac{\partial \mathbf{x}}{\partial r} \quad \text{and} \quad \mathbf{e}_2 = \frac{1}{r} \frac{\partial \mathbf{x}}{\partial \theta}$$

do not form a holonomic basis since they cannot be written as partial derivatives in any coordinate system or, equivalently,

$$\mathbf{c}_{12} = -\frac{1}{r} \mathbf{e}_2 \neq \mathbf{0}.$$

4.2 Physical and anholomic components of tensors

A second-order tensor of double field $\mathbb{T}$, whose components are referred to as anholonomic bases, can be defined as linear combinations of tensor products of the type $\mathbf{W} \otimes \mathbf{w}$ (or $\mathbf{w} \otimes \mathbf{W}$) where $\mathbf{W}$ and $\mathbf{w}$ are elements that belong to one of the bases of the sets $\{\mathbf{E}, \mathbf{E}^*, \mathbf{F}, \mathbf{F}^*\}$ and $\{\mathbf{e}, \mathbf{e}^*, \mathbf{f}, \mathbf{f}^*\}$, respectively. For instance, the expression

$$\mathbb{T} = \sum_{A,c} \left[T^{A\,c}\right]_{\mathcal{B}} \mathbf{F}_A \otimes \mathbf{e}_c, \quad (A = 1, 2, 3; c = 1, 2, 3) \tag{4.76}$$

represents a contravariant second-order tensor of double field. Here, we have that $\mathcal{B} = \mathbf{F} \otimes \mathbf{e}$ is a basis of the considered tensorial space and $[T^{A\,c}]_{\mathcal{B}}$ denotes the anholonomic components of the tensor of double field referred to the basis $\mathcal{B}$. Instead, if the components of $\mathbb{T}$ are referred to a natural basis, we have

$$\mathbb{T} = T^{I\,k} \mathbf{G}_I \otimes \mathbf{g}_k \quad (I = 1, 2, 3; k = 1, 2, 3). \tag{4.77}$$

From the condition of invariance of $\mathbb{T}$, we can write

$$\mathbb{T} = \left[T^{A\,c}\right]_{F \otimes e} \mathbf{F}_A \otimes \mathbf{e}_c = T^{Ik} \mathbf{G}_I \otimes \mathbf{g}_k \quad (A = 1, 2, 3; c = 1, 2, 3). \tag{4.78}$$

Now, taking into account the expressions of $\mathbf{F}_A = F_I^A \mathbf{G}_I$ and $\mathbf{e}_c = e_k^c \mathbf{g}_k$ as functions of $\mathbf{G}_I$ and $\mathbf{g}_i$, respectively, we obtain

$$\left[T^{A\,c}\right]_{F\otimes e} = T^{I\,k} F_I^A e_k^c, \tag{4.79}$$

where the sum is over the repeated indices. This formula is the law of transformation, under change of bases, of the contravariant components $T^{I\,k}$ of a second-order tensor of double field to their anholonomic counterpart $[T^{A\,c}]_{F\otimes e}$. Analogously, the laws of transformation of the tensor components $T_{I\,k}$, $T^{I}_{\cdot\,k}$, and $T_I^{\cdot\,k}$ to their anholonomic counterparts referred to the bases $\mathbf{F}^* \otimes \mathbf{e}^*$, $\mathbf{F} \otimes \mathbf{e}^*$, and $\mathbf{F}^* \otimes \mathbf{e}$, respectively, are given by

$$[T_{A\,c}]_{F^*\otimes e^*} = T_{Ik} F_A^I e_c^k, \tag{4.80}$$

$$\left[T^{A}_{\cdot c}\right]_{F\otimes e^*} = T^{I}_{\cdot\,k} F_I^A e_c^k, \tag{4.81}$$

$$[T_{A\cdot}^{c}]_{F^*\otimes e} = T_I^{\cdot\,k} F_A^I e_k^c. \tag{4.82}$$

Single tensor fields can be included as a special kind of double tensor fields. In fact, after swapping $\mathbf{g}$ by $\mathbf{G}$, $\mathbf{g}^*$ by $\mathbf{G}^*$, $\mathbf{e}$ by $\mathbf{E}$, $\mathbf{e}^*$ by $\mathbf{E}^*$, $\mathbf{f}$ by $\mathbf{F}$, and $\mathbf{f}^*$ by $\mathbf{F}^*$, the preceding formulas lead to the following anholonomic components of a single tensor field

$$\left[T^{AC}\right]_{F\otimes E} = T^{IK} F_I^A E_K^C, \tag{4.83}$$

$$[T_{AC}]_{F^*\otimes E^*} = T_{IK} F_A^I E_C^K, \tag{4.84}$$

$$\left[T^{A}_{\cdot C}\right]_{F\otimes E^*} = T^{I}_{K} F_I^A E_C^K, \tag{4.85}$$

$$\left[T_{A\cdot}^{C}\right]_{F^*\otimes E} = T_{I\cdot}^{K} F_A^I E_K^C, \tag{4.86}$$

which hold true for the points of the region C.

The laws of transformation under change of bases of the components of a second-order tensor of single field to their anholonomic counterparts are listed in Table 4.1.

Consider all anholonomic components of the tensor component T^{AC},

$$\left[T^{AC}\right]_{E\otimes E} = T^{IK} E_I^A E_K^C, \tag{4.87}$$

$$\left[T^{AC}\right]_{F\otimes E} = T^{IK} F_I^A E_K^C, \tag{4.88}$$

$$[T^{AC}]_{E\otimes F} = T^{IK} E_I^A F_K^C, \tag{4.89}$$

$$[T^{AC}]_{F\otimes F} = T^{IK} F_I^A F_K^C, \tag{4.90}$$

as listed in Table 4.1.

Table 4.1. Anholonomic counterparts of the components of a second-order tensor of single field.

$\left[T^{ab}\right]_{e\otimes e} = \sum_{i,j} T^{ij} e_i^a e_j^b$	$\left[T^{ab}\right]_{e\otimes f} = \sum_{i,j} T^{ij} e_i^a f_j^b$
$\left[T^{ab}\right]_{f\otimes e} = \sum_{i,j} T^{ij} f_i^a e_j^b$	$\left[T^{ab}\right]_{f\otimes f} = \sum_{i,j} T^{ij} f_i^a f_j^b$
$\left[T_a^{\cdot\, b}\right]_{e^*\otimes e} = \sum_{i,j} T_i^{\cdot\, j} e_a^i e_j^b$	$\left[T_a^{\cdot\, b}\right]_{e^*\otimes f} = \sum_{i,j} T_i^{\cdot\, j} e_a^i f_j^b$
$\left[T_a^{\cdot\, b}\right]_{f^*\otimes e} = \sum_{i,j} T_i^{\cdot\, j} f_a^i e_j^b$	$\left[T_a^{\cdot\, b}\right]_{f^*\otimes f} = \sum_{i,j} T_i^{\cdot\, j} f_a^i f_j^b$
$[T^a_{\cdot\, b}]_{e\otimes e^*} = \sum_{i,j} T^i_{\cdot\, j} e_i^a e_b^j$	$[T^a_{\cdot\, b}]_{e\otimes f^*} = \sum_{i,j} T^i_{\cdot\, j} e_i^a f_b^j$
$[T^a_{\cdot\, b}]_{f\otimes e^*} = \sum_{i,j} T^i_{\cdot\, j} f_i^a e_b^j$	$[T^a_{\cdot\, b}]_{f\otimes f^*} = \sum_{i,j} T^i_{\cdot\, j} f_i^a f_b^j$
$[T_{ab}]_{e^*\otimes e^*} = \sum_{i,j} T_{ij} e_a^i e_b^j$	$[T_{ab}]_{e^*\otimes f^*} = \sum_{i,j} T_{ij} e_a^i f_b^j$
$[T_{ab}]_{f^*\otimes e^*} = \sum_{i,j} T_{ij} f_a^i e_b^j$	$[T_{ab}]_{f^*\otimes f^*} = \sum_{i,j} T_{ij} f_a^i f_b^j$

Now we rewrite the factors of basis transformation [formulas (4.72) and (4.73)] for the undeformed configuration, that is,

$$E_K^A = \delta_K^A \, h_K \quad \text{and} \quad F_K^A = \frac{\delta_K^A}{h^K}, \tag{4.91}$$

$$F_A^K = \delta_A^K \, h^K \quad \text{and} \quad E_A^K = \frac{\delta_A^K}{h_K}, \tag{4.92}$$

with the metric symbols $h_K = \sqrt{G_{KK}} = \sqrt{\mathbf{G}_K \cdot \mathbf{G}_K}$ and $h^K = \sqrt{G^{KK}} = \sqrt{\mathbf{G}^K \cdot \mathbf{G}^K}$, which again do not contribute to the summation convention since they contain double indices.

The substitution of these factors in expressions (4.87) through (4.90) yields the following physical components :

$$\begin{aligned} \left[T^{AC}\right]_{E\otimes E} &= T^{AC} \sqrt{G_{AA}} \sqrt{G_{CC}}, \quad \left[T^{AC}\right]_{F\otimes E} = T^{AC} \frac{\sqrt{G_{CC}}}{\sqrt{G^{AA}}}, \\ \left[T^{AC}\right]_{E\otimes F} &= T^{AC} \frac{\sqrt{G_{AA}}}{\sqrt{G^{CC}}}, \\ \left[T^{AC}\right]_{E\otimes F} &= T^{AC} \frac{\sqrt{G_{AA}}}{\sqrt{G^{CC}}}, \quad \left[T^{AC}\right]_{F\otimes F} = T^{AC} \frac{1}{\sqrt{G^{AA}} \sqrt{G^{CC}}}. \end{aligned} \tag{4.93}$$

Definition 4.7 The coefficients that appear multiplying the tensor components in the five relations (4.93) above are called **physical components** of the tensor $\mathbb{T}$ relative to the bases $E \otimes E$, $F \otimes E$, $E \otimes F$, and $F \otimes F$, respectively. We use the bracket notation $[\cdot]_{\mathcal{B}}$ to denote physical components of a tensor referred to a basis $\mathcal{B}$ of the considered vector space.

Similar substitutions in the expressions of the anholonomic components listed in Table 4.1 yields the physical components listed in Table 4.2. These quantities are simpler to interpret physically than the tensor components. In [Altman and Oliveira, 1995], it was shown that they carry the same dimensions as the rectangular Cartesian components.

Table 4.2. Physical counterparts of the components of a second-order tensor of single field.

$\left[T^{ab}\right]_{e\otimes e} = T^{ab}\sqrt{g_{aa}}\sqrt{g_{bb}}$	$\left[T^{ab}\right]_{e\otimes f} = T^{ab}\sqrt{g_{aa}}/\sqrt{g^{bb}}$
$\left[T^{ab}\right]_{f\otimes e} = T^{ab}\sqrt{g_{bb}}/\sqrt{g^{aa}}$	$\left[T^{ab}\right]_{f\otimes f} = T^{ab}/\sqrt{g^{aa}}\sqrt{g^{bb}}$
$\left[T_{a}^{\cdot\,b}\right]_{e^*\otimes e} = T_{a}^{\cdot\,b}\sqrt{g_{bb}}/\sqrt{g_{aa}}$	$\left[T_{a}^{\cdot\,b}\right]_{e^*\otimes f} = T_{a}^{\cdot\,b}/\sqrt{g_{aa}}\sqrt{g^{bb}}$
$\left[T_{a}^{\cdot\,b}\right]_{f^*\otimes e} = T_{a}^{\cdot\,b}\sqrt{g^{aa}}\sqrt{g_{bb}}$	$\left[T_{a}^{\cdot\,b}\right]_{f^*\otimes f} = T_{a}^{\cdot\,b}\sqrt{g^{aa}}\sqrt{g^{bb}}$
$[T^{a}_{\cdot\,b}]_{e\otimes e^*} = T^{a}_{\cdot\,b}\sqrt{g_{aa}}/\sqrt{g_{bb}}$	$[T^{a}_{\cdot\,b}]_{e\otimes f^*} = T^{a}_{\cdot\,b}\sqrt{g_{aa}}\sqrt{g^{bb}}$
$[T^{a}_{\cdot\,b}]_{f\otimes e^*} = T^{a}_{\cdot\,b}/\sqrt{g^{aa}}\sqrt{g_{bb}}$	$[T^{a}_{\cdot\,b}]_{f\otimes f^*} = T^{a}_{\cdot\,b}\sqrt{g^{bb}}/\sqrt{g^{aa}}$
$[T_{ab}]_{e^*\otimes e^*} = T_{ab}/\sqrt{g_{aa}}\sqrt{g_{bb}}$	$[T_{ab}]_{e^*\otimes f^*} = T_{ab}\sqrt{g^{bb}}/\sqrt{g_{aa}}$
$[T_{ab}]_{f^*\otimes e^*} = T_{ab}\sqrt{g^{aa}}/\sqrt{g_{bb}}$	$[T_{ab}]_{f^*\otimes f^*} = T_{ab}\sqrt{g^{aa}}\sqrt{g^{bb}}$

Ericksen [Ericksen, 1960] already noticed that the method of anholonomic components does not include all cases, as for example, Green and Zerna's definition of physical components. [Truesdell, 1953] and [Ericksen, 1960] dealt only with anholonomic components of single field referred to the bases $\mathbf{E}$ and $\mathbf{E}^*$ of the undeformed configuration (or e or e^* of the deformed configuration) of the body and, therefore, omitted the bases $\mathbf{F}$ and $\mathbf{F}^*$ (or $\mathbf{f}$ and $\mathbf{f}^*$). The anholonomic components listed in Table 4.1, which contains Green and Zerna's formula, namely, $\left[T^{ab}\right]_{f\otimes e} = T^{ij} f_i^a e_j^b$, are 16 in number. Only four components are left after omitting the bases $\mathbf{f}$ and $\mathbf{f}^*$. This fact shows the importance of keeping the bases $\mathbf{f}$ and $\mathbf{f}^*$. The same reasoning holds for anholonomic components of higher-order tensors of single or double field. The theory here developed is more general because it deals with anholonomic components of single and double fields referred to bases whose elements belong to one of the sets $\{\mathbf{E}, \mathbf{E}^*, \mathbf{F}, \mathbf{F}^*\}$ and $\{\mathbf{e}, \mathbf{e}^*, \mathbf{f}, \mathbf{f}^*\}$. It should be noted that expressions (4.80), (4.81), and (4.82) coincide with the ones defined by [Ericksen, 1960], [Green and Zerna, 1954], and [Sedov, 1965], respectively. The theory developed here is general, since besides dealing with the components defined by Ericksen, Green, Zerna, and Sedov, it contains the definitions produced by [Synge and Schild, 1978] and [Truesdell, 1953].

Formulas (4.80), (4.81), and (4.82) can be extended to higher-order tensors of double field as follows.

Definition 4.8 The quantities

$$\left[T^{A_1 \dots A_R a_1 \dots a_r}_{B_1 \dots B_Q b_1 \dots b_q}\right]_{\mathcal{B}}$$

are said to be the **anholonomic components of the mixed double tensor field** of contravariant order R, r and covariant order Q, q referred to the basis $\mathcal{B}$ given by

$$\underbrace{E \otimes \cdots \otimes E}_{R \text{ times}} \otimes \underbrace{E^* \otimes \cdots \otimes E^*}_{Q \text{ times}} \otimes \underbrace{f \otimes \cdots \otimes f}_{r \text{ times}} \otimes \underbrace{f^* \otimes \cdots \otimes f^*}_{q \text{ times}}$$

iff $\left[T^{A_1 \dots A_R a_1 \dots a_r}_{B_1 \dots B_Q b_1 \dots b_q}\right]_{\mathcal{B}}$ is equal to

$$E^{A_1}_{I_1} \dots E^{A_R}_{I_R} E^{K_1}_{B_1} \dots E^{K_Q}_{B_Q} f^{a_1}_{i_1} \dots f^{a_r}_{i_r} f^{k_1}_{b_1} \dots f^{k_q}_{b_q} T^{I_1 \dots I_R i_1 \dots i_r}_{K_1 \dots K_Q k_1 \dots k_q}. \tag{4.94}$$

As an example, let us apply this definition to the mixed third-order tensor of double field

$$\mathbb{T} = T^{Ii}_{K} \vec{G}_I \otimes \vec{g}_i \otimes \vec{G}^K, \tag{4.95}$$

Suppose we want to find out the anholonomic components of $\mathbb{T}$ referred to the basis $\mathcal{B} = F \otimes e \otimes F^*$. Thus, from the definition we have

$$\left[T^{Aa}_{B}\right]_{F \otimes e \otimes F^*} = T^{Ii}_{K} F^A_I e^a_i F^K_B. \tag{4.96}$$

The substitution of formulas (4.72) and (4.73) into formula (4.96) leads to

$$\left[T^{Aa}_{B}\right]_{F \otimes e \otimes F^*} = \frac{\sqrt{g_{aa}}}{\sqrt{G^{AA}}} \sqrt{G^{BB}} T^{Aa}_{B}. \tag{4.97}$$

No summation over the double indices in the metric symbols is implied. This is the expression of the physical component of the third-order tensor of double field referred to the basis $\mathcal{B} = F \otimes e \otimes F^*$.

As an additional example, suppose, for instance, that the bases $\{g, g^*\}$ and $\{G, G^*\}$ change to natural and anholonomic ones. Then, the anholonomic components of the mixed fourth-order tensor components T^{Ij}_{kR}, referred to the bases $\mathcal{B} = G' \otimes e \otimes g^{*\prime} \otimes F^*$, read

$$\left[T^{A'c}_{b'D}\right]_{\mathcal{B}} = T^{Ij}_{kR} G^{A'}_I e^c_j g^k_{b'} F^R_D \tag{4.98}$$

where $G_I^{A'} = \dfrac{\partial X^{A'}}{\partial X^I}$, $g_{b'}^k = \dfrac{\partial x^k}{\partial x^{b'}}$, and e_j^c and F_R^D are (anholonomic) factors of basis transformation.

When all the bases are holonomic, formula (4.98) can be rewritten as

$$\left[T^{A'c'}_{b'D'}\right]_{\mathcal{B}'} = T^{A'c'}_{b'D'} = T^{Ij}_{kR}\, G^{A'}_I\, g^{c'}_j\, g^k_{b'}\, G^R_{D'} \tag{4.99}$$

where $\mathcal{B}' = G' \otimes g' \otimes g^{*\prime} \otimes G^{*\prime}$ and $G_I^{A'}$, $g_j^{c'}$, $g_{b'}^k$, $G_{D'}^R$ are (holonomic) factors of basis transformation. Formula (4.99) itself can be expressed in the traditional form of coordinate transformation:

$$T^{A'c'}_{b'D'} = T^{Ij}_{kR} \frac{\partial X^{A'}}{\partial X^I} \frac{\partial x^{c'}}{\partial x^j} \frac{\partial x^k}{\partial x^{b'}} \frac{\partial X^R}{\partial X^{D'}} \tag{4.100}$$

where

$$g_j^{c'} = \frac{\partial x^{c'}}{\partial x^j}, \quad G^R_{D'} = \frac{\partial X^R}{\partial X^{D'}} \tag{4.101}$$

and the (holonomic) factors of bases transformation $G_I^{A'}$ and $g_{b'}^k$ were defined above. This shows that the traditional coordinate transformation is a particular case of formula (4.98). Therefore, the change of basis allows for the unification of both the anholonomic and the natural tensor calculus in just one context.

4.3 Coordinate transformations of physical components of tensors

We will obtain here formulas for the change of physical components due to coordinate transformations. These formulas are kin to the existing ones for tensor components. In studying the tensorial fields, we are confronted with the necessity of considering an arbitrary reference frame so that we need in consequence to develop the rules for transformation of arbitrary curvilinear coordinates. Thus, let us consider the transformation from one coordinate system $\mathbf{x} = (x^1, x^2, \ldots, x^n)$ to another $\mathbf{X} = (X^1, X^2, \ldots, X^n)$ or vice versa. If the transformation

$$x^i = x^i(X^1, X^2, \ldots, X^n) \tag{4.102}$$

is such that the determinant of the Jacobian matrix does not vanish, i.e.,

$$\det J = \det \begin{bmatrix} \dfrac{\partial x^1}{\partial X^1} & \cdots & \dfrac{\partial x^1}{\partial X^n} \\ \vdots & \ddots & \vdots \\ \dfrac{\partial x^n}{\partial X^1} & \cdots & \dfrac{\partial x^n}{\partial X^n} \end{bmatrix} \neq 0,$$

then the transformation (4.102) is locally invertible and its inverse is denoted as

$$X^i = X^i(x^1, x^2, \ldots, x^n). \tag{4.103}$$

When we change coordinates from $\mathbf{x}$ to $\mathbf{X}$, the tensors transform according to the relation

$$T_{PQ}^{MN}(\mathbf{X}) = \{\det J\ (\mathbf{x}, \mathbf{X})\}^w \frac{\partial X^M}{\partial x^a}\frac{\partial X^N}{\partial x^b}\frac{\partial x^c}{\partial X^P}\frac{\partial x^d}{\partial X^Q} T_{cd}^{ab}(\mathbf{x}).$$

The quantities T_{PQ}^{MN} and T_{cd}^{ab} are the components in the $\mathbf{X}$ and $\mathbf{x}$ coordinate systems, respectively, of a tensor field of weight w of contravariant order 2, and covariant order 2, where $\det J\ (\mathbf{x}, \mathbf{X})$ is the determinant of the Jacobian matrix of the transformation from $\mathbf{x}$ to $\mathbf{X}$.

The expressions written in terms of physical components are formally analogous to the existing ones for the tensor components. For instance, we have the following example.

Example 4.3 Show that the expression written in terms of tensor components

$$T_{MR} = \frac{\partial x^m}{\partial X^M} T_{mq} \frac{\partial x^q}{\partial X^R} \tag{4.104}$$

transforms according to the following formula in terms of physical components

$$[T_{MR}]_{B^* \otimes B^*} = \left[\frac{\partial x^m}{\partial X^M}\right]_{b \otimes B^*} [T_{mq}]_{b^* \otimes b^*} \left[\frac{\partial x^q}{\partial X^R}\right]_{b \otimes B^*}. \tag{4.105}$$

Solution: Indeed taking B * = E * and b * = e *, from Table 4.2 we have

$$[T_{MR}]_{B^* \otimes B^*} = \frac{T_{MR}}{\sqrt{G_{MM}}} \frac{1}{\sqrt{G_{RR}}} \tag{4.106}$$

and

$$[T_{mq}]_{b^* \otimes b^*} = \frac{T_{mq}}{\sqrt{g_{mm}}} \frac{1}{\sqrt{g_{qq}}}. \tag{4.107}$$

Substituting the tensor components T_{MR} and T_{mq}

$$T_{MR} = \sqrt{G_{MM}}\,\sqrt{G_{RR}}\,[T_{MR}]_{B^*\otimes B^*}$$
$$T_{mq} = \sqrt{g_{mm}}\,\sqrt{g_{qq}}\,[T_{mq}]_{b^*\otimes b^*}$$

from (4.106) and (4.107) into formula (4.104), we get

$$[T_{MR}]_{B^*\otimes B^*} = \frac{\partial x^m}{\partial X^M}\frac{\sqrt{g_{mm}}}{\sqrt{G_{MM}}}[T_{mq}]_{b^*\otimes b^*}\frac{\partial x^q}{\partial X^R}\frac{\sqrt{g_{qq}}}{\sqrt{G_{RR}}}.$$

Since, again from Table 4.2,

$$\left[\frac{\partial x^m}{\partial X^M}\right]_{b\otimes B^*} = \frac{\sqrt{g_{mm}}}{\sqrt{G_{MM}}}\frac{\partial x^m}{\partial X^M} \quad \text{and} \quad \left[\frac{\partial x^q}{\partial X^R}\right]_{b\otimes B^*} = \frac{\sqrt{g_{qq}}}{\sqrt{G_{RR}}}\frac{\partial x^q}{\partial X^R},$$

we obtain (4.105).

Analogous reasoning leads to the formulas for other components, such as

$$[T_{MR}]_{B\otimes B} = \left[\frac{\partial X^M}{\partial x^m}\right]_{B\otimes b^*}[T^{mq}]_{b\otimes b}\left[\frac{\partial X^R}{\partial x^q}\right]_{B\otimes b^*}, \qquad (4.108)$$

and all such formulas can be easily generalized to tensors of higher order.

From Table 4.2 we can check that

$$[J_{a.}^{\cdot D}]_{e^*\otimes E} = \frac{\sqrt{G_{DD}}}{\sqrt{g_{aa}}}J_{a.}^{\cdot D},$$
$$[J_{\cdot a}^{D\cdot}]_{E\otimes e^*} = \frac{\sqrt{G_{DD}}}{\sqrt{g_{aa}}}J_{\cdot a}^{D\cdot}.$$

From this we see that formally (that is, regardless of the Jacobians) we have that the formulas for transformation of coordinates formally do not change if we exchange the order of the basis of the Jacobians. Note that

$$[J_{a.}^{\cdot D}]_{e^*\otimes E} = \left([J_{\cdot a}^{D\cdot}]_{E\otimes e^*}\right)^T.$$

Therefore, whenever we have a symmetry, that is, $J_{a.}^{\cdot D} = J_{\cdot a}^{D\cdot}$, we shall also obtain the equality of the physical components.

4.4 Examples of transformation of coordinates for physical components

Let X^A denote the coordinates before a transformation is introduced, which we call Lagrangian coordinates. If x^a denotes the new coordinates after transformation, which is called Eulerian coordinates, the two set of coordinates are related by

$$dx^a = \frac{\partial x^a}{\partial X^A} dX^A \tag{4.109}$$

so that we have in component form:

$$[dx^a]_e = \left[\frac{\partial x^a}{\partial X^A}\right]_{e\otimes E^*} \left[dX^A\right]_E . \tag{4.110}$$

Firstly, in order to obtain the Jacobians and vector components, we have

$$\begin{aligned}
[dx^a]_e &= e^a_k dx^k = \delta^a_k \sqrt{g_{kk}} dx^k = \sqrt{g_{aa}} dx^a, \\
\left[dX^A\right]_E &= E^A_K dX^K = \delta^A_K \sqrt{G_{KK}} dX^K = \sqrt{G_{AA}} dX^A.
\end{aligned}$$

Substituting $[dx^a]_e = \sqrt{g_{aa}} dx^a$ and $\left[dX^A\right]_E = \sqrt{G_{AA}} dX^A$ in (4.110), we get

$$\sqrt{g_{aa}} dx^a = \left[\frac{\partial x^a}{\partial X^A}\right]_{e\otimes E^*} \sqrt{G_{AA}} dX^A,$$

so that

$$dx^a = \left[\frac{\partial x^a}{\partial X^A}\right]_{e\otimes E^*} \frac{\sqrt{G_{AA}}}{\sqrt{g_{aa}}} dX^A. \tag{4.111}$$

From (4.109) and (4.111) we conclude that

$$\frac{\partial x^a}{\partial X^A} = \frac{\sqrt{G_{AA}}}{\sqrt{g_{aa}}} \left[\frac{\partial x^a}{\partial X^A}\right]_{e\otimes E^*},$$

that is,

$$\left[\frac{\partial x^a}{\partial X^A}\right]_{e\otimes E^*} = \frac{\sqrt{g_{aa}}}{\sqrt{G_{AA}}} \frac{\partial x^a}{\partial X^A},$$

which is in agreement with the definition of physical components of a double field second-order tensor **T**:

$$[T^a_{\cdot A}]_{e\otimes E^*} = \frac{\sqrt{g_{aa}}}{\sqrt{G_{AA}}} T^a_{\cdot A}. \tag{4.112}$$

See Table 4.2 or use $[T^a_{\cdot A}]_{e\otimes E^*} = e^a_i E^K_A T^i_{\cdot K}$. Thus the expression corresponding to the formula below

$$v^a = \frac{\partial x^a}{\partial X^A} V^A \tag{4.113}$$

will be

$$[v^a]_e = \left[\frac{\partial x^a}{\partial X^A}\right]_{e\otimes E^*} \left[V^A\right]_E . \tag{4.114}$$

Note that (4.113) is contained in (4.114). To see this, it suffices to take $e = g$ and $E = G$.

Thus, the anholonomic components [and, consequently, the physical components (4.114)] will **transform in the same way as the tensor components** (4.113). Clearly, as a result of symbolic precision, (4.113) can be written as (4.114):

$$[v^a]_g = \left[\frac{\partial x^a}{\partial X^A}\right]_{g\otimes G^*} [V^A]_G \tag{4.115}$$

where

$$\begin{aligned}
&[v^a]_g = v^a, \\
&\left[\frac{\partial x^a}{\partial X^A}\right]_{g\otimes G^*} = \frac{\partial x^a}{\partial X^A}, \\
&[V^A]_E = V^A.
\end{aligned}$$

Also, we can invert formula (4.115) by writing

$$[V^A]_E = \left[\frac{\partial X^A}{\partial x^a}\right]_{E\otimes e^*} [v^a]_e, \tag{4.116}$$

where

$$\left[\frac{\partial X^A}{\partial x^a}\right]_{E\otimes e^*} = \frac{\sqrt{G_{AA}}}{\sqrt{g_{aa}}} \frac{\partial x^a}{\partial X^A}. \tag{4.117}$$

In the case of changing from a basis $\mathbf{F}$ to a basis $\mathbf{e}$ we shall get the following expression

$$[v^a]_e = \left[\frac{\partial x^a}{\partial X^A}\right]_{e\otimes F^*} [V^A]_F, \tag{4.118}$$

where

$$\left[\frac{\partial x^a}{\partial X^A}\right]_{e\otimes F^*} = \sqrt{g_{aa}}\sqrt{G^{AA}} \frac{\partial x^a}{\partial X^A}.$$

Note that

$$\left[\frac{\partial x^a}{\partial X^A}\right]_{e\otimes F^*} = e_i^a F_A^K \frac{\partial x^i}{\partial X^K} = \delta_i^a \sqrt{g_{ii}} \delta_A^K \sqrt{G^{KK}} \frac{\partial x^i}{\partial X^K} = \sqrt{g_{aa}}\sqrt{G^{AA}} \frac{\partial x^a}{\partial X^A}.$$

We now proceed to other types of basis transformations.

4.4.1 Transformation of covariant components

We know that a covariant component v_a of a vector **V** will transform according to the expression

$$v_a = \frac{\partial X^D}{\partial x^a} V_D.$$

This formula can be written in terms of physical components as follows:

(1) Using bases $\mathbf{e}^*$ and $\mathbf{E}^*$ for the vector components,

$$[v_a]_{e*} = \left[\frac{\partial X^D}{\partial x^a}\right]_{e*\otimes E} [V_D]_{E*}. \tag{4.119}$$

(2) Using bases $\mathbf{e}^*$ and $\mathbf{F}^*$ for the vector components,

$$[v_a]_{e*} = \left[\frac{\partial X^D}{\partial x^a}\right]_{e*\otimes F} [V_D]_{F*}. \tag{4.120}$$

(3) Using bases $\mathbf{e}^*$ and $\mathbf{F}^*$ for the vector components,

$$[v_a]_{f*} = \left[\frac{\partial X^D}{\partial x^a}\right]_{f*\otimes E} [V_D]_{E*}. \tag{4.121}$$

(4) Using bases $\mathbf{e}^*$ and $\mathbf{F}^*$ for the vector components,

$$[v_a]_{f*} = \left[\frac{\partial X^D}{\partial x^a}\right]_{f*\otimes F} [V_D]_{F*}. \tag{4.122}$$

4.4.2 Transformation of second-order tensors

(1) For the **tensor product of two vectors $\mathbf{V}\otimes\mathbf{W}$**, we consider the product $V_D W_R$ and one can easily show that

$$\begin{aligned}[v_a w_b]_{e*\otimes e*} &= [v_a]_{e*}[w_b]_{e*} \\ &= \left[\frac{\partial X^D}{\partial x^a}\right]_{e*\otimes E}\left[\frac{\partial X^R}{\partial x^b}\right]_{e*\otimes E} [V_D]_{E*}[W_R]_{E*}.\end{aligned} \tag{4.123}$$

(2) For **twice covariant tensors**, according to (4.123) a tensor $\mathbf{T}_{DR}$ will transform as follows

$$[t_{ab}]_{e*\otimes e*} = \left[\frac{\partial X^D}{\partial x^a}\right]_{e*\otimes E}\left[\frac{\partial X^R}{\partial x^b}\right]_{E\otimes e*} [T_{DR}]_{E*\otimes E*}, \tag{4.124}$$

which is the same as

$$[t_{ab}]_{e*\otimes e*} = \left[\frac{\partial X^D}{\partial x^a}\right]_{e*\otimes E} [T_{DR}]_{E*\otimes E*} \left[\frac{\partial X^R}{\partial x^b}\right]_{E\otimes e*}. \tag{4.125}$$

This last formula exhibits the balancing of bases in a more perspicuous fashion and the two examples below demonstrate this formula.

Example 4.4 Show that (4.125) reduces to the classical formula

$$t_{ab} = \frac{\partial X^D}{\partial x^a} T_{DR} \frac{\partial X^R}{\partial x^b}. \tag{4.126}$$

Solution: To do this, we evaluate the respective physical components (taking $h_a \triangleq \sqrt{g_{aa}}$ and $H_R \triangleq \sqrt{G_{RR}}$):

$$[t_{ab}]_{e*\otimes e*} = \frac{t_{ab}}{h_a h_b},$$

$$\left[\frac{\partial X^D}{\partial x^a}\right]_{e*\otimes E} = \frac{\partial X^D}{\partial x^a}\frac{H_D}{h_a},$$

$$\left[\frac{\partial X^R}{\partial x^b}\right]_{E\otimes e*} = \frac{H_R}{h_b}\frac{\partial X^R}{\partial x^b},$$

$$[T_{DR}]_{E*\otimes E*} = \frac{T_{ab}}{H_a H_b}.$$

Substituting these physical components in (4.125) yields

$$\frac{t_{ab}}{h_a h_b} = \frac{\partial X^D}{\partial x^a}\frac{H_D}{h_a}\frac{T_{DR}}{H_D H_R}\frac{\partial X^R}{\partial x^b}\frac{H_R}{h_b} = \frac{\partial X^D}{\partial x^a} T_{DR} \frac{\frac{\partial X^R}{\partial x^b}}{h_a h_b},$$

so that

$$t_{ab} = \frac{\partial X^D}{\partial x^a} T_{DR} \frac{\partial X^R}{\partial x^b}.$$

Example 4.5 Show that the formula below reduces to the classic formula

$$[t_{ab}]_{e*\otimes f*} = \left[\frac{\partial X^D}{\partial x^a}\right]_{e*\otimes E} [T_{DR}]_{E*\otimes F*} \left[\frac{\partial X^R}{\partial x^b}\right]_{F\otimes f*}.$$

Solution: Again, to do this we evaluate the respective physical components

$$[T_{DR}]_{E*\otimes F*} = T_{DR}\frac{H^R}{H_D},$$
$$\left[\frac{\partial X^R}{\partial x^b}\right]_{F\otimes f*} = \frac{\partial X^R}{\partial x^b}\frac{h^b}{H^R},$$
$$\left[\frac{\partial X^D}{\partial x^a}\right]_{e*\otimes E} = \frac{\partial X^D}{\partial x^a}\frac{H_D}{h_a}.$$

Substituting on the right-hand side,

$$\left[\frac{\partial X^D}{\partial x^a}\right]_{e*\otimes E}[T_{DR}]_{E*\otimes F*}\left[\frac{\partial X^R}{\partial x^b}\right]_{F\otimes e*} = \frac{\partial X^D}{\partial x^a}\frac{H_D}{h_a}T_{DR}\frac{H^R}{H_D}\frac{\partial X^R}{\partial x^b}\frac{h^b}{H^R},$$
$$\left[\frac{\partial X^D}{\partial x^a}\right]_{e*\otimes E}[T_{DR}]_{E*\otimes F*}\left[\frac{\partial X^R}{\partial x^b}\right]_{F\otimes e*} = \frac{\partial X^D}{\partial x^a}T_{DR}\frac{\partial X^R}{\partial x^b}\frac{h^b}{h_a}.$$

Equating with the left-hand side gives

$$[t_{ab}]_{e*\otimes f*} = t_{ab}\frac{h^b}{h_a} = \frac{\partial X^D}{\partial x^a}T_{DR}\frac{\partial X^R}{\partial x^b}\frac{h^b}{h_a},$$

that is,

$$t_{ab} = \frac{\partial X^D}{\partial x^a}T_{DR}\frac{\partial X^R}{\partial x^b}.$$

We leave as an exercise to check the equivalence with (4.126) of the following formulas:

$$(1)\;\; [t_{ab}]_{f*\otimes f*} = \left[\frac{\partial X^D}{\partial x^a}\right]_{f*\otimes E}[T_{DR}]_{E*\otimes F*}\left[\frac{\partial X^R}{\partial x^b}\right]_{F\otimes f*}.$$

$$(2)\;\; [t_{ab}]_{f*\otimes f*} = \left[\frac{\partial X^D}{\partial x^a}\right]_{f*\otimes F}[T_{DR}]_{F*\otimes F*}\left[\frac{\partial X^R}{\partial x^b}\right]_{F\otimes f*}.$$

$$(3)\;\; [t_{ab}]_{f*\otimes e*} = \left[\frac{\partial X^D}{\partial x^a}\right]_{f*\otimes F}[T_{DR}]_{F*\otimes E*}\left[\frac{\partial X^R}{\partial x^b}\right]_{E\otimes f*}.$$

(3) For **twice covariant tensors**, we already know from the classical Tensor Calculus that a twice contravariant tensor will transform according to the expression

$$T^{MR} = \frac{\partial X^M}{\partial x^m}t^{mq}\frac{\partial X^R}{\partial x^q}. \tag{4.127}$$

Example 4.6 Referring to the method used in the two previous examples, check the equivalence with (4.127) of the following formula:

$$[T^{MR}]_{E\otimes E} = \left[\frac{\partial X^M}{\partial x^m}\right]_{E\otimes e*} [t^{mq}]_{e\otimes e} \left[\frac{\partial X^R}{\partial x^q}\right]_{e*\otimes E}.$$

Solution: Evaluating the respective physical components, we have

$$[t^{mq}]_{e\otimes e} = t^{mq} h_m h_q, \qquad \left[\frac{\partial X^M}{\partial x^m}\right]_{E\otimes e*} = \frac{\partial X^M}{\partial x^m}\frac{H_M}{h_m},$$

$$\left[\frac{\partial X^R}{\partial x^q}\right]_{e*\otimes E} = \frac{\partial X^R}{\partial x^q}\frac{H_R}{h_q}, \qquad [T^{MR}]_{E\otimes E} = T^{MR} H_M H_R.$$

Substituting on the right-hand side,

$$\left[\frac{\partial X^M}{\partial x^m}\right]_{E\otimes e*} [t^{mq}]_{e\otimes e} \left[\frac{\partial X^R}{\partial x^q}\right]_{e*\otimes E} = \frac{\partial X^M}{\partial x^m}\frac{H_M}{h_m} t^{mq} h_m h_q \frac{\partial X^R}{\partial x^q}\frac{H_R}{h_q}$$

$$= \frac{\partial X^M}{\partial x^m} H_M t^{mq} \frac{\partial X^R}{\partial x^q} H_R,$$

and equating with the left-hand side, we have

$$[T^{MR}]_{E\otimes E} = T^{MR} H_M H_R = \frac{\partial X^M}{\partial x^m} H_M t^{mq} \frac{\partial X^R}{\partial x^q} H_R.$$

Thus,

$$T^{MR} = \frac{\partial X^M}{\partial x^m} t^{mq} \frac{\partial X^R}{\partial x^q}.$$

As before, we leave as an exercise to check the equivalence with (4.127) of the following formulas:

$$(1) \quad [T^{MR}]_{E\otimes F} = \left[\frac{\partial X^M}{\partial x^m}\right]_{E\otimes e*} [t^{mq}]_{e\otimes f} \left[\frac{\partial X^R}{\partial x^q}\right]_{f*\otimes F}.$$

$$(2) \quad [T^{MR}]_{F\otimes F} = \left[\frac{\partial X^M}{\partial x^m}\right]_{F\otimes f*} [t^{mq}]_{f\otimes f} \left[\frac{\partial X^R}{\partial x^q}\right]_{f*\otimes F}.$$

$$(3) \quad [t^{mq}]_{e\otimes e} = \left[\frac{\partial x^m}{\partial X^M}\right]_{e\otimes E*} [T^{MR}]_{E\otimes E} \left[\frac{\partial x^r}{\partial X^R}\right]_{E*\otimes e}.$$

4.4.3 *Transformation of mixed components*

The mixed third-order tensor components t_n^{mq} change in accordance with the tradicional form of coordinates transformation:

$$T_N^{MR} = \frac{\partial X^M}{\partial x^m} \frac{\partial x^n}{\partial X^N} t_n^{mq} \frac{\partial X^R}{\partial x^q}.$$

The formula for the transformation from coordinate $\mathbf{x}$ to coordinate $\mathbf{X}$ that is equivalent to the one above in terms of physical components is given by

$$[T_N^{MR}]_{E\otimes F\otimes E^*} = \left[\frac{\partial X^M}{\partial x^m}\right]_{E\otimes e*} \left[\frac{\partial X^R}{\partial x^q}\right]_{f*\otimes F} [t^{mq}]_{e\otimes f\otimes e*} \left[\frac{\partial x^n}{\partial X^N}\right]_{e\otimes E^*}.$$

A tensor $\mathbb{T}$ with generic components will transform according to the following formula:

$$T_{pqr\ldots}^{abc\ldots} = N_i^a N_j^b N_k^c \ldots J_p^s J_q^t J_r^u T_{siu\ldots}^{ijk\ldots},$$

where the indices may belong either to holonomic bases or to anholonomic ones. Here, J is the second-order mixed Jacobian tensor and $N = J^{-1}$.

For example, for the case of the tensor $\mathbb{T}$ of the formula

$$[T_N^{MR}]_{E\otimes F\otimes E^*} = \left[\frac{\partial X^M}{\partial x^m}\right]_{E\otimes e*} \left[\frac{\partial X^R}{\partial x^q}\right]_{f^*\otimes F} [t_n^{mq}]_{e\otimes f\otimes e*} \left[\frac{\partial x^n}{\partial X^N}\right]_{e\otimes E^*},$$

we can identify N and J as follows

$$N_m^M = \left[\frac{\partial X^M}{\partial x^m}\right]_{E\otimes e*}, \quad N_q^R = \left[\frac{\partial X^R}{\partial x^q}\right]_{f*\otimes F}, \quad J_N^n = \left[\frac{\partial x^n}{\partial X^N}\right]_{e\otimes E^*}.$$

The correspondence basis-index is

(1) in $[T_N^{MR}]_{E\otimes F\otimes E^*}, M \to E, R \to F, N \to E^*;$

(2) in $[t_n^{mq}]_{e\otimes f\otimes e*}, m \to e, q \to f, n \to e^*$

(3) in $\left[\frac{\partial X^M}{\partial x^m}\right]_{E\otimes e*}, M \to E, m \to e^*;$

(4) in $\left[\frac{\partial X^R}{\partial x^q}\right]_{f^*\otimes F}, R \to F, q \to f^*;$

(5) in $\left[\frac{\partial x^n}{\partial X^N}\right]_{e\otimes E^*}, n \to e, N \to E^*.$

so that we can write

$$[T_N^{MR}]_{E\otimes F\otimes E^*} = N_m^M N_q^R J_N^n [t^{mq}]_{e\otimes f\otimes e*}.$$

From all these considerations, we can conclude that the physical components of tensors will transform by means of formulas that are equivalent to their corresponding ones from the classical tensor calculus.

4.4.4 *Raising and lowering indices*

We have the following rules for raising and lowering indices.

(1) *For order 1 tensors* (i.e., vectors).
Raising: Multiplying by the contravariant metric tensor (and contracting) raises an index:

$$g^{ij}A_j = A^i.$$

Lowering: Multiplying by the covariant metric tensor (and contracting) lowers an index:

$$g_{ij}A^j = A_i.$$

(2) *For order second-order tensors.*
Raising: For a second-order tensor, twice multiplying by the contravariant metric tensor and contracting in different indices raises each index:

$$T^{ab} = g^{ac}g^{bd}T_{cd}.$$

Lowering: Twice multiplying by the covariant metric tensor and contracting in different indices lowers each index:

$$T_{ab} = g_{ac}g_{bd}T^{cd}.$$

(3) *For physical components of second-order tensors.*
Raising: For a second-order tensor, twice multiplying by the contravariant metric tensor and contracting in different indices raises each index. Below we give three examples of raising indices in this case.

(a) Relating $[T_{cd}]_{e^*\otimes e^*}$ with $\left[T^{ab}\right]_{f\otimes f}$:

$$\left[T^{ab}\right]_{f\otimes f} = [g^{ac}]_{f\otimes e}\left[g^{bd}\right]_{f\otimes e}[T_{cd}]_{e^*\otimes e^*}.$$

(b) Relating $[T_{cd}]_{f^*\otimes f^*}$ with $\left[T^{ab}\right]_{e\otimes e}$:

$$\left[T^{ab}\right]_{e\otimes e} = [g^{ac}]_{e\otimes f}\left[g^{bd}\right]_{e\otimes f}[T_{cd}]_{f^*\otimes f^*}.$$

(c) Relating $[T_{cd}]_{f^*\otimes f^*}$ with $[T^a_b]_{e\otimes f^*}$:

$$[T^a_b]_{e\otimes f^*} = [g^{ac}]_{e\otimes f}[T_{cd}]_{f^*\otimes f^*}.$$

Lowering: multiplying by the bicovariant metric tensor (and contracting) lowers an index:

$$[T_a]_{e^*} = [g_{ac}]_{e^* \otimes e^*} [T^c]_e \,,$$
$$[T_a]_{e^*} = [g_{ac}]_{e^* \otimes f^*} [T^c]_f \,,$$
$$[T_a]_{f^*} = [g_{ac}]_{f^* \otimes f^*} [T^c]_f \,.$$

4.5 Anholonomic connections

4.5.1 *Frames and matrices of connections*

A system of basis vectors $\mathbf{g} = \{\mathbf{g}_i\}$, $i = 1, 2, 3$, is called a *frame* and is represented as a column vector. For typographical convenience, a column vector will usually be written as a generic vector in braces. The frames $\mathbf{e} = \{\mathbf{e}_a\}$, $\mathbf{e}^* = \{\mathbf{e}^a\}, \mathbf{f} = \{\mathbf{f}_a\}$, and $\mathbf{f}^* = \{\mathbf{f}^a\}$, $a = 1, 2, 3$, are related to the frames $\mathbf{g} = \{\mathbf{g}_i\}$ and $\mathbf{g}^* = \{\mathbf{g}^a\}$ as follows:

$$\{\mathbf{e}_a\} = E\,\{\mathbf{g}_i\} = \left[\frac{\delta^i_a}{\sqrt{g_{ii}}}\right] \{\mathbf{g}_i\}, \tag{4.128}$$

$$\{\mathbf{f}^a\} = F^*\,\{\mathbf{g}^i\} = \left[\frac{\delta^a_i}{\sqrt{g^{ii}}}\right] \{\mathbf{g}^i\}, \tag{4.129}$$

$$\{\mathbf{e}^a\} = E^*\,\{\mathbf{g}^i\} = [\delta^a_i \sqrt{g_{ii}}]\,\{\mathbf{g}^i\}, \tag{4.130}$$

$$\{\mathbf{f}_a\} = F\,\{\mathbf{g}_i\} = \left[\delta^i_a \sqrt{g^{ii}}\right] \{\mathbf{g}_i\}, \tag{4.131}$$

where the $3{\times}3$ matrices E, F, E^*, and F^* are called *matrices of transformation of frames*. Expression (4.131), for instance, can be written in the form

$$\begin{Bmatrix} \mathbf{f}_1 \\ \mathbf{f}_2 \\ \mathbf{f}_3 \end{Bmatrix} = \begin{bmatrix} h^1 & 0 & 0 \\ 0 & h^2 & 0 \\ 0 & 0 & h^3 \end{bmatrix} \begin{Bmatrix} \mathbf{g}_1 \\ \mathbf{g}_2 \\ \mathbf{g}_3 \end{Bmatrix}, \tag{4.132}$$

where $h^i = \sqrt{g^{ii}}$.

Let $\{\mathbf{b}_a\}$ be a frame related to the natural frame by a matrix of transformation J, that is,

$$\{\mathbf{b}_a\} = J\{\mathbf{g}_i\}.$$

We introduce the *Cartan matrix* $C(J)$ of a differentiable nonsingular square matrix J by defining (see Cartan[Cartan, 1983])

$$C(J) = (dJ).J^{-1}, \tag{4.133}$$

with dJ being the derivative of J. We also introduce the following notation:

$$C(J)_a^b = \frac{dh^a}{h^b}\delta_a^b, \tag{4.134}$$

$$J_a^i = \delta_a^i h^i, \tag{4.135}$$

$$N_i^b = \frac{1}{h^i}\delta_i^b, \tag{4.136}$$

where $N = J^{-1}$.

Example 4.7 Consider, from (4.132), the matrices

$$J = \begin{bmatrix} h^1 & 0 & 0 \\ 0 & h^2 & 0 \\ 0 & 0 & h^3 \end{bmatrix} \tag{4.137}$$

and its inverse

$$J^{-1} = \begin{bmatrix} \frac{1}{h^1} & 0 & 0 \\ 0 & \frac{1}{h^2} & 0 \\ 0 & 0 & \frac{1}{h^3} \end{bmatrix}. \tag{4.138}$$

Find the Cartan matrix $C(J)$ of the differentiable nonsingular square matrix J.

Solution: We have

$$(dJ)\cdot J^{-1} = \begin{bmatrix} dh^1 & 0 & 0 \\ 0 & dh^2 & 0 \\ 0 & 0 & dh^3 \end{bmatrix} \begin{bmatrix} \frac{1}{h^1} & 0 & 0 \\ 0 & \frac{1}{h^2} & 0 \\ 0 & 0 & \frac{1}{h^3} \end{bmatrix} = \begin{bmatrix} \frac{dh^1}{h^1} & 0 & 0 \\ 0 & \frac{dh^2}{h^2} & 0 \\ 0 & 0 & \frac{dh^3}{h^3} \end{bmatrix}. \tag{4.139}$$

Consider the associated matrices Γ, Ω, and Λ obtained by representing the derivatives of the vectors of the natural and the anholonomic bases,

$$d\{\mathbf{g}_i\} = \Gamma\{\mathbf{g}_j\}, \tag{4.140}$$

$$d\{\mathbf{e}_a\} = \Omega\{\mathbf{e}_b\}, \tag{4.141}$$

$$d\{\mathbf{f}_a\} = \Lambda\{\mathbf{f}_b\}. \tag{4.142}$$

The matrix Γ is called a *natural connection* while Λ and Ω are called *anholomic connections*. We should observe that the matrices Γ, Ω, and Λ are functions of the frames (as opposed to points) referred to bases of the type $\mathbf{b} \otimes \mathbf{b}^*$ (for instance Λ is a matrix referred to the bases $\mathbf{f} \otimes \mathbf{f}^*$). The following definition summarizes the formulas (4.140), (4.141), and (4.142).

Definition 4.9 Let $\{\mathbf{b}_a\}$ be a frame such that $\{\mathbf{b}_a\} = J\{\mathbf{g}_i\}$. A matrix Ψ is said to be a **matrix of a connection relative to** J if

$$d\{\mathbf{b}_a\} = \Psi\{\mathbf{b}_b\}. \tag{4.143}$$

The next theorem provides us with an expression to calculate Ψ in a function of Γ.

Theorem 4.1 *Let Γ be a natural connection. If Ψ is a matrix of connection relative to J, then*

$$\Psi = J\Gamma J^{-1} + C(J),$$

where $C(J) = (dJ)\, J^{-1}$ is the Cartan matrix of J.

■ **Proof:** We have seen that $\{\mathbf{b}_a\} = J\{\mathbf{g}_i\}$, so that

$$d\{\mathbf{b}_a\} = dJ\{\mathbf{g}_j\} + Jd\{\mathbf{g}_j\} \tag{4.144}$$

$$= dJ\{\mathbf{g}_j\} + J\Gamma\{\mathbf{g}_j\} \tag{4.145}$$

$$= (dJ + J\Gamma)\{\mathbf{g}_j\}. \tag{4.146}$$

Since

$$d\{\mathbf{b}_a\} = \Psi\{\mathbf{b}_b\} \tag{4.147}$$

$$= \Psi J\{\mathbf{g}_j\}, \tag{4.148}$$

we have that

$$\Psi J\{\mathbf{g}_j\} = (dJ + J\Gamma)\{\mathbf{g}_j\}, \tag{4.149}$$

which implies

$$\Psi = C(J) + J\Gamma J^{-1}. \qquad \Box \tag{4.150}$$

One can check that, in terms of components, the above formula reads

$$\Psi^b_{\cdot a} = J^i_a \Gamma^j_i N^b_j + d(J^i_a) N^b_i \tag{4.151}$$

$$= J^i_a \Gamma^j_i N^b_j + (C(J))^b_{\cdot a}, \tag{4.152}$$

where $N = J^{-1}$. Also, in the case of the connection Λ we have

$$\Lambda^b_{\cdot a} = \frac{\delta^a_i h^i}{\delta^b_j h^j} \Gamma^j_i + \frac{d(h^a)}{h^b} \delta^b_a \tag{4.153}$$

$$= \frac{h^a}{h^b} \Gamma^b_a + \frac{d(h^a)}{h^b} \delta^b_a, \tag{4.154}$$

where $J = F$ and $h^i = \sqrt{g^{ii}}$.

The quantities Ψ^b_{ac} defined by

$$\Psi^b_{ac} = \mathbf{b}^b \cdot \partial_c \mathbf{b}_a, \tag{4.155}$$

where $\partial_c = \frac{\partial}{\partial x^c}$, are called connection symbols of Ψ. Let us observe that this definition is equivalent to

$$\partial_c \mathbf{b}_a = \Psi^b_{ac} \mathbf{b}_b. \tag{4.156}$$

Since our reference is the natural basis, we can write, in terms of the bracket notation,

$$\Psi^b_{\cdot ac} = [\Gamma^b_{ac}]_{b \otimes b^* \otimes b^*}, \tag{4.157}$$

that is, $\Psi^b_{\cdot ac}$ are the anholonomic components of Γ^b_{ac}. Note that $\mathbf{b}$ stands for either $\mathbf{e}$, $\mathbf{e}^*$, $\mathbf{f}$, or $\mathbf{f}^*$. For instance, from the formula

$$\Lambda^b_{\cdot ac} = [\Gamma^b_{ac}]_{f \otimes f^* \otimes f^*} \tag{4.158}$$

we can deduce that

$$\Lambda^b_{\cdot ac} = [\Gamma^b_{ac}]_{f \otimes f^* \otimes f^*} = \frac{h^a h^c}{h^b} \Gamma^b_{ac} + \frac{(h^a)_{,c}\, h^c}{h^b} \delta^b_a. \tag{4.159}$$

There are several kinds of anholonomic components of Γ. For instance,

$$\Omega^b_{\cdot ac} = [\Gamma^b_{ac}]_{e \otimes e^* \otimes e^*} = \frac{h_b}{h_a h_c} \Gamma^b_{ac} + h_b \frac{(1/h_a)_{,c}}{h_c} \delta^b_a, \tag{4.160}$$

$$[\Gamma^b_{ac}]_{f \otimes e^* \otimes e^*} = \frac{1}{h_a h_c h^b} \Gamma^b_{ac} + \frac{(1/h_\alpha)_{,c}}{h^b h_c} \delta^b_a. \tag{4.161}$$

4.5.2 *More on Christoffel symbols*

In order to get some feeling about what the Christoffel symbols mean, let us take a differentiable curve $\mathbf{r} = \mathbf{r}(t)$ through a point r_o of the body. The tangent vector $d\mathbf{r}$ to this curve can be written in terms of curvilinear coordinates x^i in each of the following representations

$$d\mathbf{r} = dx^i \mathbf{g}_i, \tag{4.162}$$

$$d\mathbf{r} = [dx^a]_e \, \mathbf{e}_a, \tag{4.163}$$

$$d\mathbf{r} = [dx^a]_f \, \mathbf{f}_a. \tag{4.164}$$

It is clear that the vectors $\mathbf{g}_j = \dfrac{\partial}{\partial x^j}\mathbf{r} \triangleq \partial_j \mathbf{r}$ are tangent to the coordinate curves of x^j. The vectors $\{\mathbf{g}_j\}$ form the so called natural (or holonomic) basis and associated to them we have the reciprocal basis $\{\mathbf{g}^j\}$ defined by $\mathbf{g}^j \cdot \mathbf{g}_i = \delta^j_i$. Also, the vectors $\partial_j \mathbf{g}_i$ themselves can be represented in terms of the basis $\{\mathbf{g}_k\}$. The Christoffel symbols Γ^k_{ji} are the coefficients of this expansion:

$$\partial_j \mathbf{g}_i = \Gamma^k_{ji} \mathbf{g}_k. \tag{4.165}$$

Using the reciprocal basis on (4.165), we can calculate them to give

$$\Gamma^k_{ji} = \mathbf{g}^k \cdot \partial_j \mathbf{g}_i, \tag{4.166}$$

or, more explicitly,

$$\Gamma^k_{ji} = \mathbf{g}^k \cdot \frac{\partial \mathbf{g}_i}{\partial x^j}. \tag{4.167}$$

On the other hand, given an anholonomic basis $\{\mathbf{b}_a\}$, we can introduce the *generalized Christoffel symbols* Ψ^b_{ca} by defining them as (see [Schouten, 1951], formula IV 7.7)

$$\Psi^b_{ca} = \mathbf{b}^b \cdot \partial_c \mathbf{b}_a, \tag{4.168}$$

where ∂_c denotes the derivative

$$\partial_c = J^k_c \frac{\partial}{\partial x^k}. \tag{4.169}$$

and

$$J^k_c = \mathbf{b}_a \cdot \mathbf{g}^k. \tag{4.170}$$

These generalized Christoffel symbols, which we also refer to as *anholonomic connections*, are split up into two cases (and, correspondingly, two notations), depending upon the choice for one of the anholonomic bases, $\mathbf{e}$ or $\mathbf{f}$. Thus, we define

$$\Omega^b_{ca} = \mathbf{e}^b \cdot \partial_c \mathbf{e}_a, \tag{4.171}$$

and

$$\Lambda^b_{ca} = \mathbf{f}^b \cdot \partial_c \mathbf{f}_a. \tag{4.172}$$

The derivative symbol ∂_c has a different meaning here for each one of the formulas. In the case of physical components in terms of the unitary basis

$$\mathbf{e}_a = \frac{1}{h^a}\mathbf{g}_a, \tag{4.173}$$

it means

$$\partial_c = \frac{1}{h_c}\frac{\partial}{\partial x^c}.$$

Indeed, by (4.73) we can write

$$\partial_c = e^k_c \frac{\partial}{\partial x^k} = \frac{1}{h_k}\delta^k_c \frac{\partial}{\partial x^k} = \frac{1}{h_c}\frac{\partial}{\partial x^c}.$$

Similarly, if we take

$$\mathbf{f}_a = h^a \mathbf{g}_a, \tag{4.174}$$

which are the reciprocal counterpart of the unit basis

$$\mathbf{f}^a = \frac{1}{h^a}\mathbf{g}^a, \tag{4.175}$$

the meaning of ∂_c is

$$\partial_c = f^k_c \frac{\partial}{\partial x^k} = \delta^k_c \, h^k \frac{\partial}{\partial x^k} = h^c \frac{\partial}{\partial x^c}.$$

Recall (4.24), (4.25), (4.73), and (4.26). Thus, the anholonomic connections (4.171) and (4.172) become

$$\Omega^b_{ca} = \mathbf{e}^b \cdot \frac{1}{h_c}\frac{\partial \mathbf{e}_a}{\partial x^c} \tag{4.176}$$

$$\Lambda^b_{ca} = \mathbf{f}^b \cdot h^c \frac{\partial \mathbf{f}_a}{\partial x^c}. \tag{4.177}$$

The table below summarizes these representations of the Christoffel symbols:

Basis	Matrices of connection	Connection symbol	Derivation formula
Natural $\mathbf{g} \otimes \mathbf{g}^*$	Γ	Γ^b_{ca}	$\Gamma^b_{ca} = \mathbf{g}^b \cdot \dfrac{\partial \mathbf{g}_a}{\partial x^c}$
Anholonomic $\mathbf{e} \otimes \mathbf{e}^*$	Ψ	Ω^b_{ca}	$\Omega^b_{ca} = \mathbf{e}^b \cdot \dfrac{1}{h_c} \dfrac{\partial \mathbf{e}_a}{\partial x^c}$
Anholonomic $\mathbf{f} \otimes \mathbf{f}^*$	Ψ	Λ^b_{ca}	$\Lambda^b_{ca} = \mathbf{f}^b \cdot h^c \dfrac{\partial \mathbf{f}_a}{\partial x^c}$

The following result gives formulas for the anholonomic connections in terms of physical components.

Proposition 4.1 *The anholonomic connections are given in terms of the metric symbols by:*

$$\Omega^b_{ca} = \delta^b_a \frac{h_b}{h_c} \left(\frac{\partial}{\partial x^c} \left(\frac{1}{h_b} \right) \right) + \frac{h_b}{h_c h_a} \Gamma^b_{ca}, \tag{4.178}$$

$$\Lambda^b_{ca} = \delta^b_a \frac{h^c}{h^b} \frac{\partial h^b}{\partial x^c} + \frac{h^c h^a}{h^b} \Gamma^b_{ca}. \tag{4.179}$$

■ **Proof:** To show the formula for Ω^b_{ca}, let us first evaluate de derivative of the unitary vectors

$$\mathbf{e}_a = \frac{1}{h_a} \mathbf{g}_a. \tag{4.180}$$

The product rule gives

$$\begin{aligned}
\frac{\partial \mathbf{e}_a}{\partial x^c} &= \frac{\partial}{\partial x^c} \left(\frac{1}{h_a} \mathbf{g}_a \right) = \left[\frac{\partial}{\partial x^c} \left(\frac{1}{h_a} \right) \right] \mathbf{g}_a + \frac{1}{h_a} \frac{\partial \mathbf{g}_a}{\partial x^c} \\
&= \left[\frac{\partial}{\partial x^c} \left(\frac{1}{h_a} \right) \right] \mathbf{g}_a + \frac{1}{h_a} \Gamma^b_{ca} \mathbf{g}_b \\
&= \left[\frac{\partial}{\partial x^c} \left(\frac{1}{h_a} \right) \right] h_a \mathbf{e}_a + \frac{1}{h_a} \Gamma^b_{ca} h_b \mathbf{e}_b,
\end{aligned}$$

where we have used the definition of the Christoffel symbols as the coefficients of the expansion of the derivative of the natural basis in terms of the vectors of the natural basis, that is, from (4.165),

$$\frac{\partial \mathbf{g}_a}{\partial x^c} = \Gamma^b_{ca} \mathbf{g}_b,$$

and we have used the representation for $\mathbf{g}_i$ from (4.180). Therefore,

$$\begin{aligned}
\Omega^b_{\ ca} &= \mathbf{e}^b \cdot \frac{1}{h_c}\frac{\partial \mathbf{e}_a}{\partial x^c} \\
&= \mathbf{e}^b \cdot \left\{ \frac{1}{h_c}\left[\frac{\partial}{\partial x^c}\left(\frac{1}{h_a}\right)\right] h_a \mathbf{e}_a + \frac{1}{h_c}\frac{1}{h_a}\Gamma^b_{\ ca} h_b \mathbf{e}_b \right\} \\
&= \frac{1}{h_c}\left[\frac{\partial}{\partial x^c}\left(\frac{1}{h_a}\right)\right] h_a\ \mathbf{e}^b \cdot \mathbf{e}_a + \frac{1}{h_c}\frac{1}{h_a}\Gamma^b_{\ ca} h_b \mathbf{e}^b \cdot \mathbf{e}_b \\
&= \frac{1}{h_c}\left(\frac{\partial}{\partial x^c}\left(\frac{1}{h_a}\right)\right) h_a\ \delta^b_a + \frac{h_b}{h_c h_a}\Gamma^b_{\ ca}
\end{aligned}$$

and (4.178) follows.

To show the formula for $\Lambda^b_{\ ca}$, first note that from (4.174) we can express the derivative as

$$\begin{aligned}
\frac{\partial \mathbf{f}_a}{\partial x^c} &= \frac{\partial}{\partial x^c}\left(h^a \mathbf{g}_a\right) = \frac{\partial h^a}{\partial x^c}\mathbf{g}_a + h^a \frac{\partial \mathbf{g}_a}{\partial x^c} \\
&= \frac{\partial h^a}{\partial x^c}\mathbf{g}_a + h^a \Gamma^b_{\ ca}\mathbf{g}_b \\
&= \frac{\partial h^a}{\partial x^c}\frac{1}{h^a}\mathbf{f}_a + h^a \Gamma^b_{\ ca}\frac{1}{h^b}\mathbf{f}_b.
\end{aligned}$$

Hence,

$$\begin{aligned}
\Lambda^b_{\ ca} &= \mathbf{f}^b \cdot h^c \frac{\partial \mathbf{f}_a}{\partial x^c} \\
&= \mathbf{f}^b \cdot h^c \left(\frac{\partial h^a}{\partial x^c}\frac{1}{h^a}\mathbf{f}_a + h^a \Gamma^b_{\ ca}\frac{1}{h^b}\mathbf{f}_b\right) \\
&= \frac{h^c}{h^a}\frac{\partial h^a}{\partial x^c}\mathbf{f}^b \cdot \mathbf{f}_a + \frac{h^a h^c}{h^b}\Gamma^b_{\ ca}\mathbf{f}^b \cdot \mathbf{f}_b \\
&= \delta^b_a \frac{h^c}{h^b}\frac{\partial h^b}{\partial x^c} + \frac{h^a h^c}{h^b}\Gamma^b_{\ ca} \quad \square.
\end{aligned}$$

4.5.3 *Anholonomic covariant derivatives of tensors*

Let $\mathbf{w} = \mathsf{w}^a \mathbf{b}_a$ be a vector field, where we are using the notation sans serif $\mathsf{w}^a = [w^a]_b$ and $\mathsf{w}_a = [w_a]_{b^*}$ to stand for the anholonomic components of $\mathbf{w}$.

For each connection Ψ there are two kinds of components of $\dfrac{\partial \mathbf{w}}{\partial x^c}$, called *anholonomic covariant derivatives* of $\mathbf{w}$, which are defined as follows:

$$\frac{\partial \mathbf{w}}{\partial x^c} = (\mathsf{w}^b_{;c})\mathbf{b}_b \quad \text{if and only if} \quad \mathsf{w}^b_{;c} = \frac{\partial \mathbf{w}}{\partial x^c} \cdot \mathbf{b}^b \tag{4.181}$$

and

$$\frac{\partial \mathbf{w}}{\partial x^c} = (\mathsf{w}_{b;c})\mathbf{b}^b \quad \text{if and only if} \quad \mathsf{w}_{b;c} = \frac{\partial \mathbf{w}}{\partial x^c} \cdot \mathbf{b}_b. \tag{4.182}$$

The following proposition is useful for the computation of the anholonomic covariant derivatives of $\mathbf{w}$.

Proposition 4.2 *Let* $\partial_c \mathbf{b}_a = \Psi^b_{ac}\mathbf{b}_b$. *Then*

$$\mathsf{w}^b_{;c} = \mathsf{w}^b_{,c} + \mathsf{w}^a \Psi^b_{ac}. \tag{4.183}$$

■ **Proof:** First, we note that

$$\partial_c \mathbf{w} = \partial_c(\mathsf{w}^a \mathbf{b}_a) = \mathbf{b}_a \partial_c(\mathsf{w}^a) + \mathsf{w}^a \partial_c \mathbf{b}_a. \tag{4.184}$$

But

$$\mathbf{b}_b \Psi^b_{ac} = \partial_c \mathbf{b}_a, \tag{4.185}$$

Therefore

$$\partial_c \mathbf{w} = \partial_c(\mathsf{w}^a \mathbf{b}_a) = \mathbf{b}_a \partial_c(\mathsf{w}^a) + \mathsf{w}^a \mathbf{b}_b \Psi^b_{ac} \tag{4.186}$$

$$= \mathbf{b}_b \partial^b_c \mathsf{w} + \mathsf{w}^a \mathbf{b}_b \Psi^b_{ac} = (\partial^b_c \mathsf{w} + \mathsf{w}^a \Psi^b_{ac})\mathbf{b}_b. \tag{4.187}$$

That is,

$$\frac{\partial \mathbf{w}}{\partial x^c} = (\partial^b_c \mathsf{w} + \mathsf{w}^a \Psi^b_{ac})\mathbf{b}_b. \tag{4.188}$$

This implies

$$\mathsf{w}^b_{;c} = \partial^b_c \mathsf{w} + \mathsf{w}^a \Psi^b_{ac} \quad \square. \tag{4.189}$$

We note that similar results hold for $\mathsf{w}_{b;c}$, that is,

$$\mathsf{w}_{b;c} = \mathsf{w}_{b,c} - \mathsf{w}_a \Psi^a_{bc}, \tag{4.190}$$

and for tensors of all orders. For instance, if we consider the component T^a_b of a mixed second-order tensor, we have

$$\mathsf{T}^a_{b;c} = \mathsf{T}^a_{b,c} + \mathsf{T}^d_b \Psi^a_{dc} - \mathsf{T}^a_d \Psi^d_{bc}. \tag{4.191}$$

We can express the anholonomic components of the covariant derivative of a tensor directly in terms of the derivatives of the anholonomic components themselves.

We shall examine one example applying directly the definition of covariant derivative of the physical components of a contravariant vector single field. For instance, let us obtain the covariant derivative of the physical components of a contravariant vector $\mathbf{U} = U^1\mathbf{g}_1 + U^2\mathbf{g}_2 + U^3\mathbf{g}_3$. To do this, take

$$\mathbf{U} = \left[U^a_{.}\right]_e \mathbf{e}_a = U^i \mathbf{g}_i,$$

where the U^i are the components of $\mathbf{U}$ with respect to the natural basis $\mathbf{g}_i$ and the $\left[U^a_{.}\right]_e$ are the physical components of $\mathbf{U}$ with respect to the anholonomic basis $\mathbf{e}$.

The components of the covariant derivatives of $\mathbf{U}$ with respect to the natural basis $\mathbf{g}$, denoted as $U^i_{;j}$, can be evaluated by means of

$$U^i_{;j} = \frac{\partial V^i}{\partial x^j} + \Gamma^i_{kj} U^k.$$

On the other hand, the covariant derivatives of $\mathbf{U}$ with respect to the basis $\mathbf{e}$ are indicated as $\left[U^a_{;b}\right]_{e\otimes e*}$ and, as we already know, are given by

$$\left[U^a_{;b}\right]_{e\otimes e*} = \frac{1}{h_b}\frac{\partial \left[U_{;a}\right]_e}{\partial x^b} + \Omega^a_{cb}\left[U^c_{.}\right]_e,$$

where

$$\Omega^a_{bc} = [\Gamma^a_{bc}]_{e\otimes e^*\otimes e^*} = \frac{h_a}{h_b h_c}\left(\Gamma^a_{bc}\right) + \frac{h_a(1/h_b)_{,c}}{h_c}\delta^a_b.$$

Substituting the above expression by on the expression for $\left[U^a_{;b}\right]_{e\otimes e*}$, we find

$$\left[U^a_{;b}\right]_{e\otimes e*} = \frac{1}{h_b}\frac{\partial\left(\left[U^a_{.}\right]_e\right)}{\partial x^b} + \left(\frac{h_a}{h_b h_c}\left(\Gamma^a_{bc}\right) + \frac{h_a(1/h_b)_{,c}}{h_c}\delta^a_b\right)\left[U^c_{.}\right]_e.$$

Thus, for all $a \neq b$ we have that $\delta^a_b = 0$ and

$$\left[U^a_{;b}\right]_{e\otimes e*} = \frac{1}{h_b}\frac{\partial\left(\left[U^a_{.}\right]_e\right)}{\partial x^b} + \frac{h_a}{h_b h_c}\Gamma^a_{bc}\left[U^c_{.}\right]_e.$$

Similarly, for all $a = b$ we have $\delta^a_b = 1$ and

$$\left[U^a_{;a}\right]_{e\otimes e*} = \frac{1}{h_a}\frac{\partial\left(\left[U^a_{.}\right]_e\right)}{\partial x^a} + \left(\frac{1}{h_c}\left(\Gamma^a_{ac}\right) + \frac{h_a(1/h_a)_{,c}}{h_c}\right)\left[U^c_{.}\right]_e. \quad \square$$

4.6 Strain tensor

We define a symmetric tensor E_{IJ}, called the strain tensor, by the expression

$$E_{IJ} = \frac{1}{2}\left(g_{IJ} - G_{IJ}\right) \tag{4.192}$$

where $\mathbf{C} = g_{IJ}\mathbf{G}^I \otimes \mathbf{G}^J = C_{IJ}\mathbf{G}^I \otimes \mathbf{G}^J = \mathbf{J}^T \cdot \mathbf{J}$ is the right Cauchy–Green deformation tensor and the G_{IJ} are the metrics in the reference configuration of a point of the body.

The strain tensor E_{IJ} may be expressed in terms of the displacement vector $\mathbf{u}$, or in its components with respect to the base vectors $\mathbf{G}_I$. In terms of displacement, the expression (4.192) becomes

$$E_{IJ} = \frac{1}{2}\left(\mathbf{G}_J \cdot \mathbf{u}_{,I} + \mathbf{G}_I \cdot \mathbf{u}_{,J} + \mathbf{u}_{,I} \cdot \mathbf{u}_{,J}\right). \tag{4.193}$$

The displacement vector $\mathbf{u}$ and its derivative $\mathbf{u}_{,I}$ may be expressed in terms of the base vectors of the reference configuration of the body $\mathcal{C}$. Thus,

$$\mathbf{u} = U_M \mathbf{G}^M, \quad \mathbf{u}_{,I} = U_{M;I}\mathbf{G}^M, \tag{4.194}$$

and the covariant differentiation, reads

$$U_{M;I} = U_{M,I} - U_R \Gamma^R_{MI}. \tag{4.195}$$

Substituting formulas (4.194) into (4.193) and using the following relations

$$\mathbf{G}_I = \mathbf{R}_{,I}, \quad \mathbf{G}^I \cdot \mathbf{G}_J = \delta^I_J, \quad G_{IJ} = \mathbf{G}_I \cdot \mathbf{G}_J \quad G^{IJ} = \mathbf{G}^I \cdot \mathbf{G}^J \tag{4.196}$$

we obtain the expression of Green's strain tensor components

$$E_{IJ} = E_{JI} = \mathring{E}_{IJ} + \frac{1}{2}U^M_{;I}U_{M;J} \quad (I, J, M = I, II, III), \tag{4.197}$$

where

$$\mathring{E}_{IJ} = \mathring{E}_{JI} = \frac{1}{2}(U_{I;J} + U_{J;I}).$$

Using the bracket notation, we write

$$[E_{AB}]_{E^*\otimes E^*} = \left[\mathring{E}_{AB}\right]_{E^*\otimes E^*} + \frac{1}{2}\left[U^C_{;A}\right]_{E\otimes E^*}[U_{C;B}]_{E^*\otimes E^*},$$

where

$$\left[\mathring{E}_{AB}\right]_{E^*\otimes E^*} = \frac{1}{2}\left([U_{I;J}]_{E^*\otimes E^*} + [U_{J;I}]_{E^*\otimes E^*}\right),$$

for $A, B = I, II, III$.

4.7 Dimensional analysis for tensors

Physical situations can be described by considering certain scalar quantities (or properties) as basic and building up a system of compound properties constructed in terms of the basic ones. The dimensional analysis is a technique used in the physical sciences to reduce physical properties, such as (the magnitude of) velocity, to its basic quantities, such as length and time, and facilitates the study of interrelationships of systems and their properties and avoids the inconvenience of incompatible measurement units. Taking the basic quantities to be length (L), time (T), mass (M), electric charge (Q), and absolute temperature (Θ) and considering them as *fundamental dimensions*, we introduce the concept of *dimension* (or physical dimension) *of physical scalar quantities* in the following way. First, to each one of the basic quantities we define its dimension to be L, T, M, Q and Θ, respectively. Then, given a compound property F (that is, F stands for a physical property defined in terms of the fundamental dimensions), we define the *dimension of* F, denoted $Dim(F)$, to be the product $L^m T^n M^p Q^q \Theta^r$ of rational powers of the dimensions of the basic properties according to how they are expressed in terms of the fundamental dimensions. For instance, since velocity is defined in terms of length divided by time, the dimension of (the magnitude of) velocity is LT^{-1}.

We say that a physical property F is *dimensionally mensurable* whenever it is described in terms of the fundamental dimensions. If two properties F_1 and F_2 are such that $Dim(F_1) = Dim(F_2)$, we say that they are *commensurable* or *physically homogeneous*. For physical properties which are not described by the basic quantities we define its dimension to be 1 so that to every physical property we shall associate a dimension. We should mention that in dimensional analysis this sets up a multiplication operation that allows one to show that the set of dimensions of physical quantities with this operation forms a multiplicative group.

We have the following **Principle of Commensurable Quantities (PCQ)**, which is actually a modified version of a theorem whose original demonstration is often attributed to the American physicist Edgar Buckingham:

> If a quantity (property) F is expressed as the (algebraic or vector) sum of other mensurable quantities $Q_1, \ldots, Q_n$, then all the quantities Q_i are commensurable and
>
> $$Dim(F) = Dim(Q_i).$$

On the other hand, the physical components of a tensor $\mathbb{T}$ are those that result from some measuring processes done in laboratories. These measurements imply that to any given tensor $\mathbb{T}$ it is associated some measure unity. Therefore, we must consider that, in addition to mathematical attributes such as holonomy or anholonomy, a tensor should have physical attributes provided by a dimensional analysis. These physical attributes can be expressed as the dimension (physical dimension) of the tensor. Thus, there are two ways for obtaining the *dimension* $Dim(\mathbb{T})$ *of a tensor* $\mathbb{T}$:

(1) evaluating its magnitude $|\mathbb{T}|$ and defining the dimension of $\mathbb{T}$ to be the dimension of the scalar quantity $|\mathbb{T}|$;

or

(2) obtaining its components $[\mathbb{T}_{ijk\cdots n}]_{\mathcal{I}}$ in an orthonormal Cartesian basis $\mathcal{I}$ and, supported by the principle of commensurable quantities, defining the dimension of $\mathbb{T}$ to be the dimension of any one of the components of $\mathbb{T}$, regardless of the sign.

Thus, the physical dimension of a tensor $\mathbb{T}$ is given by

$$Dim(\mathbb{T}) = Dim(|\mathbb{T}|) = Dim\left([\mathbb{T}_{ijk\cdots n}]_{\mathcal{I}}\right) = L^p T^q M^r Q^s \Theta^h,$$

where the exponents p, q, r, s, h are rational numbers.

For example, take the case $\mathbb{T} = a\mathbf{i} + b\mathbf{j} + c\mathbf{k}$with $\mathbf{i}, \mathbf{j}, \mathbf{k}$ being the basis in a Cartesian system. Since

$$\begin{aligned} Dim(a\mathbf{i}) &= Dim(|a\mathbf{i}|) = Dim(|a|), \\ Dim(b\mathbf{j}) &= Dim(|b|), \\ Dim(c\mathbf{k}) &= Dim(|c|), \end{aligned}$$

we have that

$$Dim(\mathbb{T}) = Dim(|a|) = Dim(|b|) = Dim(|c|),$$

which means that the dimension of $\mathbb{T}$ coincides with the dimension that is common to all its components in the Cartesian system regardless of the sign.

As it was evident from this example (for, say, $a = 1$), the dimension of a unit vector is equal to 1. More precisely, let $\mathbf{b} = (\mathbf{b}_1, \mathbf{b}_2, \cdots, \mathbf{b}_n)$ be any basis in an inner product space and consider the unit vectors

$$\mathbf{u}_a = \frac{1}{|\mathbf{b}_a|}\mathbf{b}_a, \quad a = 1, 2, \ldots, n.$$

Then, since

$$Dim(\mathbf{u}_a) = Dim\left(\left|\frac{1}{|\mathbf{b}_a|}\mathbf{b}_a\right|\right) = Dim\left(\frac{|\mathbf{b}_a|}{|\mathbf{b}_a|}\right) = Dim(1) = 1,$$

we have that

$$Dim(\mathbf{u}_a) = Dim(\mathbf{u}^a) = 1.$$

We end this section with a number of examples in which we evaluate the dimension of some specific tensors that appear in some studies in physical sciences.

Example 4.8

1. Velocity vector in polar coordinates. Consider the unitary polar basis $\mathcal{P} = (\mathbf{u}_r, \mathbf{u}_\theta)$, where $\mathbf{u}_r$ is the unit vector in the radial directon and $\mathbf{u}_\theta$ is the unit vector orthogonal to the radial direction. The *velocity* $\mathbf{v}$ of the center of gravity $(r(t), \theta(t))$ of an object moving in space is written in terms of this polar basis as

$$\mathbf{v} = \left(\frac{dr}{dt}\right)\mathbf{u}_r + \left(r\frac{d\theta}{dt}\right)\mathbf{u}_\theta.$$

Then,

$$Dim(\mathbf{v}) = Dim\left(\frac{dr}{dt}\right) = Dim\left(r\frac{d\theta}{dt}\right) = LT^{-1}.$$

Note that there is a dimensional homogeneity in the sense that the physical dimension of the vector is equal to the physical dimensions of its components. In this case, we already know that these components are the physical components of $\mathbf{v}$ in the basis $\mathcal{P}$.

2. Inertia tensor. In an orthonormal Cartesian system of coordinates, consider a rectangular solid box $\mathcal{S}$ with constant volumetric density ρ and total mass m and assume further that its sides ℓ_1, ℓ_2, ℓ_3 are parallel to the coordinate axes. Then the *inertia tensor* $\mathbf{I}$ of the solid is defined by

$$\mathbf{I} = \frac{m}{12}\left[(\ell_2^2 + \ell_3^2)\mathbf{j}_1 \otimes \mathbf{j}_1 + (\ell_1^2 + \ell_3^2)\mathbf{j}_2 \otimes \mathbf{j}_2 + (\ell_1^2 + \ell_2^2)\mathbf{j}_3 \otimes \mathbf{j}_3\right].$$

Since the products $\mathbf{j}_1 \otimes \mathbf{j}_1$, $\mathbf{j}_2 \otimes \mathbf{j}_2$ and $\mathbf{j}_3 \otimes \mathbf{j}_3$ are dimensionless, it follows that

$$\begin{aligned} Dim(\mathbf{I}) &= Dim\left(\frac{m}{12}(\ell_2^2 + \ell_3^2)\right) \\ &= Dim\left(\frac{m}{12}(\ell_2^2 + \ell_3^2)\right) \\ &= Dim\left(\frac{m}{12}(\ell_2^2 + \ell_3^2)\right) \\ &= ML^2. \end{aligned}$$

It is worthwhile to note that the dimension of tensor products of unitary bases $(\mathbf{u}_a)$ are dimensionless because

$$Dim(\mathbf{u}_a \otimes \mathbf{u}_b) = Dim(\mathbf{u}_a) \times Dim(\mathbf{u}_b) = 1 \times 1 = 1.$$

In general, we have

$$Dim(\mathbf{u}_a \otimes \mathbf{u}_b \otimes \mathbf{u}_c \cdots \otimes \mathbf{u}_n) = 1,$$
$$Dim(\mathbf{u}^a \otimes \mathbf{u}^b \otimes \mathbf{u}^c \cdots \otimes \mathbf{u}^n) = 1.$$

3. Metric tensor. Let x^i and x^j be two generic coordinates belonging to a curvilinear coordinate system C and suppose that the metric tensor $\mathbf{g}$ has coordinates g_{ij}. Considering that in curvilinear coordinates the element of arc length ds is given by $ds^2 = dx^i dx^j g_{ij}$ with $Dim(ds^2) = L^2$, it follows that

$$L^2 = Dim(ds^2) = Dim(x^i) \times Dim(x^j) \times Dim(g_{ij}).$$

Thus, the physical dimension of the metric tensor is given by

$$Dim(g_{ij}) = \frac{L^2}{Dim(x^i) \times Dim(x^j)}. \tag{4.198}$$

Let us consider the specific example of the paraboloidal system of coordinates (u, v, θ). We have that it is related to the Cartesian system (x, y, z) by the formulas

$$x = uv\cos\theta,$$
$$y = uv\sin\theta,$$
$$z = \frac{1}{2}(u^2 - v^2).$$

Then, $Dim(u) = Dim(v) = L^{1/2}$ and $Dim(\theta) = 1$ and from (4.198) we have that

$$Dim(g_{11}) = \frac{L^2}{L} = L,$$
$$Dim(g_{22}) = \frac{L^2}{L} = L,$$
$$Dim(g_{33}) = \frac{L^2}{1} = L^2.$$

It can be shown that

$$g_{11} = g_{22} = u^2 + v^2,$$
$$g_{33} = (uv)^2.$$

As for the physical dimensions of the vectors $\mathbf{g}_i$ of a natural basis, similarly to what we have done above for tensors (after all, vectors are no more than tensors of order 1), we have that

$$Dim(\mathbf{g}_i) = Dim(|\mathbf{g}_i|) = Dim(\sqrt{g_{ii}}) = \sqrt{\frac{L^2}{X^i X^i}} = \frac{L}{X^i}$$

where $X^i = Dim(x^i)$.

4. Physical components of g_{ij}. We know that the physical component of g_{ij} in terms of the basis $\mathbf{e}^* \otimes \mathbf{e}^*$ is given by

$$[g_{ij}]_{e^* \otimes e^*} = \frac{1}{\sqrt{g_{ii}}\sqrt{g_{jj}}} g_{ij}.$$

Therefore,

$$\begin{aligned} Dim\left([g_{ij}]_{e^* \otimes e^*}\right) &= \frac{1}{Dim\left(\sqrt{g_{ii}}\right) Dim\left(\sqrt{g_{jj}}\right)} Dim\left(g_{ij}\right) \\ &= \frac{1}{\frac{L}{X^i}\frac{L}{X^j}} \frac{L^2}{X^i X^j} \\ &= 1 \end{aligned}$$

so that $[g_{ij}]_{e^* \otimes e^*}$ is dimensionless.

5. Velocity in generalized coordinates. As we know, at any position P in a path C the vector velocity $\mathbf{v}$ is tangent to the path. Suppose the point P moves from P_1 to P_2 and let $\Delta\mathbf{r} = \overrightarrow{P_2 - P_1}$ at a time increment Δt. Its average velocity over the time interval is $\Delta\mathbf{r}/\Delta t$. When Δt approaches zero, we write

$$\mathbf{v} = \frac{d\mathbf{r}}{dt} = \lim_{\Delta t \to 0} \frac{\Delta\mathbf{r}}{\Delta t}.$$

In a path $x^k = x^k(q^1, q^2, \ldots, q^n, t)$, we have

$$\frac{d\mathbf{r}_k}{dt} = \frac{\partial \mathbf{r}_k}{\partial q^i}\frac{\partial q^i}{\partial t} + \frac{\partial \mathbf{r}_\mathbf{k}}{\partial t}.$$

6. Curvature and torsion of space curves. Consider a unit speed space curve $\beta : [a, b] \to E^3$ parametrized by the arc length s. We write

$$\dot{\beta} = \frac{d\beta}{ds}$$

and define the curvature of β to be the norm of its acceleration

$$\kappa = \left| \frac{d^2\beta}{ds^2} \right| = \ddot{\beta}.$$

By differentiating $\dot{\beta} \cdot \dot{\beta} = 1$ we obtain that $\dot{\beta} \cdot \ddot{\beta} = 0$, which shows that the velocity vector is perpendicular to the acceleration. Hence, if we take their cross product we get a third vector which is perpendicular to both velocity and acceleration. In this way, three mutually perpendicular unit vectors $\{\mathbf{T}(s), \mathbf{N}(s), \mathbf{B}(s)\}$ can be defined at each point of the curve β with curvature κ, namely,

$$\mathbf{T} = \dot{\beta}, \quad \mathbf{N} = \frac{1}{\kappa}\dot{\mathbf{T}}, \quad \text{and} \quad \mathbf{B} = \mathbf{T} \times \mathbf{N}.$$

Since these three vector functions form a basis, their derivatives can be represented in terms of them, giving the celebrated Frenet–Serret equations for unit speed curves:

$$\begin{aligned}
\dot{\mathbf{T}} &= \kappa \mathbf{N}, \\
\dot{\mathbf{N}} &= -\kappa \mathbf{T} + \tau \mathbf{B}, \\
\dot{\mathbf{B}} &= -\tau \mathbf{N}.
\end{aligned}$$

$\mathbf{N}$ is called the principal normal, $\mathbf{B}$ is the binormal, and τ is the torsion. We say that $\{\mathbf{T}, \mathbf{N}, \mathbf{B}\}$ is the Frenet frame field along β, consisting of three mutually perpendicular unit vectors moving along the curve with $\mathbf{T}$ always pointing forward.

Let us proceed to obtain the physical dimensions of curvature and torsion. Noting that the triad of vectors consists of unitary ones so that $Dim(\mathbf{T}) = Dim(\mathbf{N}) = Dim(\mathbf{B}) = 1$, we have that

$$\begin{aligned}
Dim(\dot{\mathbf{T}}) &= Dim(\kappa \mathbf{N}) = Dim(\kappa) \\
Dim(\dot{\mathbf{N}}) &= -Dim(\kappa \mathbf{T}) = Dim(\tau \mathbf{B}) = Dim(\kappa) = Dim(\tau) \\
Dim(\dot{\mathbf{B}}) &= -Dim(\tau \mathbf{N}) = Dim(\tau).
\end{aligned}$$

But, since

$$\begin{aligned}
Dim(\dot{\mathbf{T}}) &= Dim\left(\frac{d\mathbf{T}}{ds}\right) = L^{-1} \\
Dim(\dot{\mathbf{B}}) &= Dim\left(\frac{d\mathbf{B}}{ds}\right) = L^{-1},
\end{aligned}$$

it follows that

$$Dim(\kappa) = Dim(\tau) = L^{-1}.$$

7. Christoffel symbol. Given a sufficiently smooth natural basis $(\mathbf{g}_i)$ and its reciprocal $(\mathbf{g}^i)$, we have defined the Christoffel symbol Γ^k_{ij} of the second kind as

$$\Gamma^k_{ij} = \mathbf{g}^k \cdot \partial_i(\mathbf{g}_j).$$

Let (x^1, x^2, x^3) be coordinates in a curvilinear system of coordinates whose metric tensor $\mathbf{g}$ has components g_{ij}. If $Dim(x^i) = X^i$, then

$$Dim(\Gamma^k_{ij}) = \frac{X^k}{X^i X^j}.$$

Indeed,

$$\begin{aligned} Dim(\Gamma^k_{ij}) &= Dim\left(\mathbf{g}^k \cdot \partial_i(\mathbf{g}_j)\right) \\ &= Dim(\mathbf{g}^k).Dim(\partial_i(\mathbf{g}_j)) \\ &= \frac{X^k}{L}\frac{1}{X^j}\left(\frac{L}{X^i}\right) \\ &= \frac{X^k}{X^i X^j}. \end{aligned}$$

With the notation $|\mathbf{g}_k| = h_k$ and $|\mathbf{g}^k| = h^k$, introduced by Gabriel Lamé (1795-1870), we note that

$$Dim(h_k) = Dim(\mathbf{g}_k) = \frac{L}{X^k}$$

and

$$Dim(h^k) = \frac{1}{Dim(h_k)}.$$

Now consider a vector $\mathbf{A} = A^k \mathbf{g}_k$ and its four distinct types of physical components:

$$\begin{aligned} \left[A^k\right]_e &= h_k A^k, \\ \left[A_k\right]_{f*} &= h^k A_k, \\ \left[A^k\right]_f &= \frac{A^k}{h^k}, \\ \left[A_k\right]_{e*} &= \frac{A_k}{h_k}. \end{aligned}$$

Then,

$$Dim(\mathbf{A}) = Dim(h_k A^k) = Dim(h^k A_k) = Dim\left(\frac{A^k}{h^k}\right) = Dim\left(\frac{A_k}{h_k}\right).$$

To prove this, consider $\mathbf{A} = A^k \mathbf{g}_k$. We have that

$$Dim(\mathbf{A}) = Dim(A^k \mathbf{g}_k) = Dim(A^k h_k).$$

Thus,

$$Dim(A^k \mathbf{g}_k) = Dim(A^k h_k) = Dim\left(\left[A^k\right]_e\right), \qquad (4.199)$$

where we note that in $A^k h_k$ we do not sum in the repeated index k. From (4.199) it follows that

$$Dim(A^k \mathbf{g}_k) = Dim(h_k A^k) = Dim(h^k A_k) = Dim\left(\frac{A^k}{h^k}\right) = Dim\left(\frac{A_k}{h_k}\right).$$

Therefore, for $\mathbf{A} = A^k \mathbf{g}_k$ we have:

$$Dim(\mathbf{A}) = Dim\left[A^k\right]_e = Dim\left[A_k\right]_{f^*} = Dim\left[A^k\right]_f = Dim\left[A_k\right]_{e^*}.$$

4.7.1 *Expressions in terms of physical components*

It is possible to have tensor expressions written in terms of natural bases passed directly to expressions in terms of physical components. The following examples show this process.

Example 4.9
The expression

$$\mathbf{C} \cdot \mathbf{D} = g^{ij} C_i D_j$$

can be written as

$$\mathbf{C} \cdot \mathbf{D} = [g^{ij}]_{e \otimes e} [C_i]_{e^*} [D_j]_{e^*},$$

or

$$\mathbf{C} \cdot \mathbf{D} = [g^{ij}]_{f \otimes e} [C_i]_{f^*} . [D_j]_{e^*}.$$

We have the equivalences

$$\begin{aligned}
\mathbf{A}\cdot\mathbf{C} = A^k C_k &\iff \mathbf{A}\cdot\mathbf{C} = [A^k]_e.[C_k]_{e^*} \\
F^k = ma^k &\iff [F^k]_e = m[a^k]_e \\
A_k T^{kms} = P^{ms} &\iff [A_k]_{e^*}.[T^{kms}]_{e\otimes e\otimes e} = [P^{ms}]_{e\otimes e} \\
\mathbf{A}\otimes\mathbf{C} = A^k C_s \mathbf{g}_k \otimes \mathbf{g}^s &\iff \mathbf{A}\otimes\mathbf{C} = [A^k C_s]_{e\otimes e^*} \mathbf{e}_k \otimes \mathbf{e}^s.
\end{aligned}$$

One can also use other bases for these same expressions. For instance,

$$A_k T^{kms} = P^{ms} \iff [A_k]_{f^*}.[T^{kms}]_{f\otimes e\otimes e} = [P^{ms}]_{e\otimes e}.$$

In the following example, $\mathbb{T}$ is the *stress energy tensor* of a Newtonian fluid with velocity $\mathbf{u}$, specific weight ρ, pressure p and metric tensor $\mathbf{g}$. The components with Greek indices are components in a natural basis whereas those with indices $a, b, c, \ldots$ are the components in an anholonomic basis.

Example 4.10

1. Stress energy tensor. The stress energy tensor $\mathbb{T} = p\mathbf{g} + (p+\rho)\mathbf{u}\otimes\mathbf{u}$ can be expressed in a coordinate basis as

$$T_{\mu\nu} = pg_{\mu\nu} + (p+\rho)u_\mu u_\nu$$

and in a noncoordinate basis as

$$T_{ab} = pg_{ab} + (p+\rho)u_a u_b.$$

Since g_{ab} and $u_a \otimes u_b$ are dimensionless, we have that

$$Dim(\mathbf{T}) = Dim(p) = Dim(\rho) = \frac{MLT^{-2}}{L^3}.$$

2. The Lagrangian and the Principle of Least Action. Newton's laws are relationships among vectors, which is why they get so messy when we change coordinate systems. On the other hand, the Lagrangian formulation uses scalars only so that coordinate transformations tend to be much easier. A Lagrangian is a function of the space location of a particle and of its velocity. Given a Lagrangian $\mathcal{L}(x,t)$, we define the action:

$$S = \int_a^b \mathcal{L}(x,t)\, dt.$$

Given particular starting and ending positions, the system follows a path between the start and end points which minimizes the action.

Let $(q_1, \cdots, q_n)$ denote the space coordinates. It can be shown that, on the path of least action, we must have

$$\mathcal{E} = \frac{\partial \mathcal{L}}{\partial q_i} - \frac{\partial}{\partial t}\frac{\partial \mathcal{L}}{\partial q_i} = 0. \tag{4.200}$$

This is known as the *Euler–Lagrange equation* [Reddy, 2002]. The Lagrangian $\mathcal{L}$ is chosen in such a manner that the path of least action will be the path along which Newton's laws are obeyed. For Newtonian mechanics, the Lagrangian is chosen to be

$$\mathcal{L} = \mathcal{K} - \mathcal{V},$$

where $\mathcal{K}$ is the kinetic energy and $\mathcal{V}$ is the potential energy. In mechanics of particles, we define the kinetic energy and the i-component of generalized force $\mathcal{E}^i$ as follows:

$$\mathcal{K} = \frac{m}{2} g_{jk}\dot{q}^j\dot{q}^k, \tag{4.201}$$

$$\mathcal{E}^i = m(\ddot{q}^i + \Gamma^i_{jk}\dot{q}^j\dot{q}^k). \tag{4.202}$$

In the following two examples we deal with the dimension of the generalized force $\mathcal{E}^i$ in two situations.

Example 4.11 In the case of a spherical coordinate system (r, θ, ϕ),

$$\begin{aligned}
&ds^2 = dr^2 + r^2 d\theta^2 + r^2 \sin^2\theta\ d\phi^2,\\
&g_{11} = 1, \quad g_{22} = r^2, \quad g_{33} = r^2 \sin^2\theta,\\
&g^{11} = 1, \quad g^{22} = \frac{1}{r^2}, \quad g^{33} = \frac{1}{r^2 \sin^2\theta},\\
&q^1 = r, q^2 = \theta, q^3 = \phi,
\end{aligned}$$

obtain the physical dimension of generalized force.

Solution: Since $\mathcal{E}^i = m(\ddot{q}^i + \Gamma^i_{jk}\dot{q}^j\dot{q}^k)$ is composed from adding two terms, we must check the physical dimension of each one:

$$Dim(m(\ddot{q}^i)) = MX^iT^{-2}, \quad \text{where } X^i = Dim(q^i),$$

and

$$Dim(\Gamma^i_{jk}\dot{q}^j\dot{q}^k) = \frac{MX^i}{X^jX^k}\frac{X^jX^k}{T^2} = MX^iT^{-2}.$$

Therefore,

$$Dim\left(\mathcal{E}^i\right) = MX^iT^{-2}.$$

For $i = 1$, $q^1 = r$ and $X^1 = L$, so that $Dim\left(\mathcal{E}^1\right) = MLT^{-2}$, which is indeed the dimension of force.
For $i = 2$, $q^2 = \theta$ and $X^2 = 1$, so that $Dim\left(\mathcal{E}^2\right) = MT^{-2}$, which is not the dimension of force.
For $i = 3$, $q^3 = \phi$ and $X^3 = 1$, so that $Dim\left(\mathcal{E}^3\right) = MT^{-2}$, which is not the dimension of force.

Thus, we conclude that the generalized force has not always the physical dimension of force. This can be fixed by using the physical components in the basis **e**. Indeed,

$$\left[\mathcal{E}^2\right]_e = h_2\mathcal{E}^2 = r\mathcal{E}^2 \text{ so that } Dim(\left[\mathcal{E}^2\right]_e) = MLT^{-2} \text{ and}$$

$$\left[\mathcal{E}^3\right]_e = h_3\mathcal{E}^2 = (r\sin\theta)\,\mathcal{E}^3 \text{ so that } Dim(\left[\mathcal{E}^3\right]_e) = MLT^{-2}.$$

As for the kinetic energy $\mathcal{K}$, note that from (4.201) we can write the kinetic energy of a particle in the following matrix form:

$$\mathcal{K} = \frac{m}{2}\,(\dot{q})^T\,\mathbf{G}\mathbf{q},$$

where

$$\mathbf{q} = \begin{Bmatrix} \dot{q}^1 \\ \dot{q}^2 \\ \dot{q}^3 \end{Bmatrix}$$

and **G** is the matrix made up of the g_{jk}'s, that is,

$$\mathcal{K} = \frac{m}{2}\,\{\dot{r}\;\dot{\theta}\;\dot{\phi}\}\left(\begin{bmatrix} 1 & 0 & 0 \\ 0 & r^2 & 0 \\ 0 & 0 & r^2\sin^2\theta \end{bmatrix}\begin{Bmatrix} \dot{r} \\ \dot{\theta} \\ \dot{\phi} \end{Bmatrix}\right).$$

Since

$$\begin{bmatrix} 1 & 0 & 0 \\ 0 & r^2 & 0 \\ 0 & 0 & r^2\sin^2\theta \end{bmatrix}\begin{Bmatrix} \dot{r} \\ \dot{\theta} \\ \dot{\phi} \end{Bmatrix} = \begin{Bmatrix} \dot{r} \\ r^2\dot{\theta} \\ \frac{1}{2}r^2\dot{\phi}\,(2\sin^2\theta) \end{Bmatrix},$$

we have

$$\mathcal{K} = \frac{m}{2}\,\{\dot{r}\;\dot{\theta}\;\dot{\phi}\}\begin{Bmatrix} \dot{r} \\ r^2\dot{\theta} \\ \frac{1}{2}r^2\dot{\phi}\,(2\sin^2\theta) \end{Bmatrix}$$

so that

$$\mathcal{K} = \frac{1}{2}m\left[(\dot{r})^2 + \sin^2\theta\, r^2(\dot{\phi})^2 + r^2(\dot{\theta})^2\right]. \tag{4.203}$$

One can easily see that $\mathcal{K}$ is *commensurable*, that is, every term has the same dimension $\dfrac{ML^2}{T^2}$.

Example 4.12 In the spherical system (r, θ, ϕ), the **Euler–Lagrange covector** $\mathcal{E}$ is given by

$$\mathcal{E} = \mathcal{E}_r \mathbf{g}^r + \mathcal{E}_\theta \mathbf{g}^\theta + \mathcal{E}_\phi \mathbf{g}^\phi \tag{4.204}$$

in the natural cobasis $\mathbf{g}^*$ and

$$\mathcal{E} = [\mathcal{E}_r]_{e^*} \mathbf{e}^r + [\mathcal{E}_\theta]_{e^*} \mathbf{e}^\theta + [\mathcal{E}_\phi]_{e^*} \mathbf{e}^\phi \tag{4.205}$$

in the anholonomic cobasis $\mathbf{e}^*$. When the component $\mathcal{E}_2 = \mathcal{E}_\theta$, the potential energy V equals zero and the basis vector is $\mathbf{g}^2$. Obtain the dimensional homogeneity for $[\mathcal{E}_\theta]_{e^*}$.

Solution: According to Eq. (4.200),

$$\mathcal{E}_\theta = \frac{\partial}{\partial t}\left(\frac{\partial K}{\partial \theta}\right) - \frac{\partial K}{\partial \theta}. \tag{4.206}$$

Inserting in Eq. (4.206) the value of $\mathcal{K}$ given by Eq. (4.203) yields

$$\mathcal{E}_\theta = m\left(r^2\ddot{\theta} + 2r\dot{r}\dot{\theta} - r^2\dot{\phi}^2 \sin\theta\cos\theta\right). \tag{4.207}$$

Doing a dimensional analysis, we notice that the right-hand side of Eq. (4.207) has the following dimensions:

$$Dim(mr^2\ddot{\theta}) = M(L/T^{-1})^2 = \frac{ML^2}{T^2},$$

$$Dim(m2r\dot{r}\dot{\theta}) = \frac{ML^2}{T^2},$$

$$Dim(r^2\dot{\phi}^2\sin\theta\cos\theta) = \frac{ML^2}{T^2}.$$

Thus, the right-hand side of Eq. (4.207) is *dimensionally homogeneous* but it does not have the dimension of force. However, considering the physical components in the basis $\mathbf{e}^*$, this incompatibility dissolves itself. Thus, since $h_\theta = r$, we have that

$$[\mathcal{E}_\theta]_{e^*} = \frac{1}{h_2}\mathcal{E}_\theta = \frac{1}{h_\theta}\mathcal{E}_\theta = \frac{1}{r}\mathcal{E}_\theta.$$

Therefore,

$$[\mathcal{E}_\theta]_{e^*} = \frac{1}{r}m\left(r^2\ddot{\theta} + 2r\dot{r}\dot{\theta} - r^2\dot{\phi}^2\sin\theta\cos\theta\right) \tag{4.208}$$

and we get the dimensional homogeneity, since now the physical dimension of the second term of Eq. (4.208) turns out to be

$$Dim([\mathcal{E}_\theta]_{e^*}) = \frac{ML}{T^2},$$

which is the dimension of force.

4.8 Exercises

(1) Suppose that the transformation

$$\overline{x} = x \cdot x,$$
$$\overline{y} = y \cdot y,$$

connects the coordinate systems $(x^i) = (x, y)$ and $(\overline{x}^i) = (\bar{x}, \bar{y})$.
(a) Compute the Jacobian matrix of the transformation as well as its inverse matrix.
(b) Calculate the vectors $\overline{\mathbf{g}}_1$ and $\overline{\mathbf{g}}_2$.
Answers:

$$\text{(a)} \quad J = \begin{bmatrix} 2x & 0 \\ 0 & 2y \end{bmatrix} \quad \text{and} \quad J^{-1} = \frac{1}{2}\begin{bmatrix} \frac{1}{x} & 0 \\ 0 & \frac{1}{y} \end{bmatrix}$$

$$\text{(b)} \quad \overline{\mathbf{g}}_1 = \frac{1}{2x}\mathbf{i} \quad , \quad \overline{\mathbf{g}}_2 = \frac{1}{2y}\mathbf{j}$$

(2) The spherical coordinates (x^i) are connected to rectangular coordinates $(\overline{x}^i)$ via

$$\overline{x}^1 = x^1 \sin x^2 \cos x^3,$$
$$\overline{x}^2 = x^1 \sin x^2 \sin x^3,$$
$$\overline{x}^3 = x^1 \cos x^2.$$

Find:
(a) The natural basis of the spherical coordinates.
(b) The Jacobian matrix

$$J = \left[\frac{\partial \overline{x}}{\partial x}\right]_{3\times 3}.$$

(c) The metric coefficients in matrix form, i.e., $G = J^T J$.

Answers:

(a)

$$\begin{aligned}\mathbf{g}_1 &= \sin x^2 \cos x^3 \mathbf{i} + \sin x^2 \sin x^3 \mathbf{j} + \cos x^2 \mathbf{k},\\ \mathbf{g}_2 &= x^1(\cos x^2 \cos x^3 \mathbf{i} + \cos x^2 \sin x^3 \mathbf{j} - \sin x^2 \mathbf{k}),\\ \mathbf{g}_3 &= x^1 \sin x^2(-\sin x^3 \mathbf{i} + \cos x^3 \mathbf{j}).\end{aligned}$$

(b)

$$J = \begin{bmatrix} \sin x^2 \cos x^3 & x^1 \cos x^2 \cos x^3 & -x^1 \sin x^2 \sin x^3 \\ \sin x^2 \sin x^3 & x^1 \cos x^2 \sin x^3 & -x^1 \sin x^2 \cos x^3 \\ \cos x^2 & -x^1 \sin x^2 & 0 \end{bmatrix}.$$

(c)

$$G = J^T J = \begin{bmatrix} 1 & 0 & 0 \\ 0 & (x^1)^2 & 0 \\ 0 & 0 & (x^1 \sin x^2)^2 \end{bmatrix},$$

that is, $g_{11} = 1, g_{22} = (x^1)^2$, and $g_{33} = (x^1 \sin x^2)^2$.

(3) The parabolic coordinates (u, v, ϕ) are connected to rectangular orthonormal Cartesian coordinates $(\bar{x}^i) = (x, y, z)$ by

$$x = uv \cos\phi,$$
$$y = uv \sin\phi,$$
$$z = \frac{1}{2}(v^2 - u^2).$$

Find the metric coefficients g_{ij}.

Answer: $g_{ij} = 0$, if $i \neq j$, and $g_{11} = g_{22} = (u^2+v^2)^2$, and $g_{33} = (u^2v^2)^2$.

(4) Calculate the volume element dV corresponding to the parabolic coordinates (u, v, ϕ) given by

$$x = uv \cos\phi,$$
$$y = uv \sin\phi,$$
$$z = \frac{1}{2}(v^2 - u^2).$$

Answer: $dV = (uv\sqrt{u^2 + v^2})dudvd\phi$.

(5) The toroidal coordinates (r, θ, ϕ) are connected to rectangular orthonormal Cartesian coordinates $(\bar{x}^i) = (x, y, z)$ by

$$x = (a - r\cos\theta)\cos\phi,$$
$$y = (a - r\cos\theta)\sin\phi,$$
$$z = r\sin\phi,$$

where $a > 0$ is a given parameter. Find the metric coefficients g_{ij}. **Answer**: $g_{ij} = 0$, if $i \neq j$, and $g_{11} = 1, g_{22} = r^2$, and $g_{33} = (a - r\cos\theta)^2$.

(6) Let $\{\mathbf{i}, \mathbf{j}, \mathbf{k}\}$ be a right-handed orthonormal basis. Show that, for

$$\mathbf{A} = A_1\mathbf{i} + A_2\mathbf{j} + A_3\mathbf{k},$$
$$\mathbf{B} = B_1\mathbf{i} + B_2\mathbf{j} + B_3\mathbf{k},$$

we have that $\mathbf{A} \times \mathbf{B} = C_1\mathbf{i} + C_2\mathbf{j} + C_3\mathbf{k}$, where

$$\begin{Bmatrix} C_1 \\ C_2 \\ C_3 \end{Bmatrix} = \begin{bmatrix} 0 & -A_3 & A_2 \\ A_3 & 0 & -A_1 \\ -A_2 & A_1 & 0 \end{bmatrix} \begin{Bmatrix} B_1 \\ B_2 \\ B_3 \end{Bmatrix}.$$

(7) Show that

$$(\mathbf{a} \cdot \mathbf{b} \times \mathbf{c})(\mathbf{d} \cdot \mathbf{e} \times \mathbf{f}) = \begin{bmatrix} \mathbf{a}\cdot\mathbf{d} & \mathbf{a}\cdot\mathbf{e} & \mathbf{a}\cdot\mathbf{f} \\ \mathbf{b}\cdot\mathbf{d} & \mathbf{b}\cdot\mathbf{e} & \mathbf{b}\cdot\mathbf{f} \\ \mathbf{c}\cdot\mathbf{d} & \mathbf{c}\cdot\mathbf{e} & \mathbf{c}\cdot\mathbf{f} \end{bmatrix}.$$

(8) Compute the volume element dV corresponding to the prolate spheroidal coordinates (u, v, ϕ) given by

$$x = a\sinh y \sin v \cos\phi,$$
$$y = a\sinh u \sin v \sin\phi,$$
$$z = a\cosh u \cos v.$$

Answer: $dV = \left(a^2\sqrt{\sinh^2 u + \sin^2 v}\right) \sinh u \sin v du dv d\phi$.

(9) Let $\mathbf{T} \in V \otimes W$, $\mathbf{T} \neq \mathbf{0}$. Show that in general there exist several representations of the form

$$\mathbf{T} = \mathbf{v}_i \otimes \mathbf{w}_i, \quad \mathbf{v}_i \in V, \ \mathbf{w}_i \in W$$

with $\{\mathbf{v}_1, \ldots, \mathbf{v}_n\}$ and $\{\mathbf{w}_1, \ldots, \mathbf{w}_n\}$ each linearly independent.

(10) Let $\mathbf{a} \in V$ and $\mathbf{b} \in W$ be such that $\mathbf{b} \otimes \mathbf{a} \neq \mathbf{0}$. Show that

$$\mathbf{b} \otimes \mathbf{a} = \mathbf{w} \otimes \mathbf{v} \quad \text{if and only if} \quad \mathbf{v} = \lambda \mathbf{a} \text{ and } \mathbf{w} = \lambda \mathbf{b}, \quad \text{for some } \lambda \neq 0.$$

(11) Show that $L(V, L(V, V))$ is isomorphic to $T_2^1(V)$.

(12) Show that $g_{ij}\mathbf{A}^i\mathbf{B}^j$ is invariant, i.e., $g_{\bar{i}\,\bar{j}}\mathbf{A}^{\bar{i}}\mathbf{B}^{\bar{j}} = g_{ij}\mathbf{A}^i\mathbf{B}^j$.

(13) Associated with the vectors $\mathbf{a} = a_i\mathbf{e}^i, \mathbf{b} = b_j\mathbf{e}^j$, and $\mathbf{c} = c_k\mathbf{e}^k$, it is possible to define the following trilinear functional $\mathbf{T}$ by putting

$$\mathbf{T}(\mathbf{a}, \mathbf{b}, \mathbf{c}) = a_i b_j c_k T_k^{ij}.$$

Show that this functional defines a tensor.

(14) Associated with the standard basis $\{\mathbf{e}_1, \mathbf{e}_2, \mathbf{e}_3\}$ of $\mathbb{R}^3$,

$$\mathbf{e}_1 = \begin{Bmatrix} 1 \\ 0 \\ 0 \end{Bmatrix}, \quad \mathbf{e}_2 = \begin{Bmatrix} 0 \\ 1 \\ 0 \end{Bmatrix}, \quad \mathbf{e}_3 = \begin{Bmatrix} 0 \\ 0 \\ 1 \end{Bmatrix},$$

there corresponds a set of scalars defined by $T^i_{jk} = T(\mathbf{e}^i, \mathbf{e}_j, \mathbf{e}_k)$ where T is a given function and $\mathbf{e}^i$ is the reciprocal (defined by $\mathbf{e}^i \cdot \mathbf{e}_j = \delta^i_j$). Suppose that another basis is defined as

$$\begin{aligned} \mathbf{e}_{\bar{1}} &= \mathbf{e}_1 - 2\mathbf{e}_2, \\ \mathbf{e}_{\bar{2}} &= 2\mathbf{e}_1 + \mathbf{e}_2, \\ \mathbf{e}_{\bar{3}} &= \mathbf{e}_1 + \mathbf{e}_2. \end{aligned}$$

Find $T^{\bar{i}}_{\bar{j}\bar{k}} = T(\mathbf{e}^{\bar{i}}, \mathbf{e}_{\bar{j}}, \mathbf{e}_{\bar{k}})$ in terms of T^i_{jk}.

(15) Show that $T^i_j\mathbf{v}_i\mathbf{w}^j$ is invariant.

(16) Show that if $a_j = T_{ij}\mathbf{v}^i$ are the components of a covariant vector $\mathbf{a}$ for all contravariant vectors $\mathbf{v}$, then T_{ij} is a covariant tensor of order 2.

(17) If v^i are the components of a contravariant vector, when can you say that a^i defined by $a^i = v^i + \alpha v_i$, $\alpha \in \mathbb{R}$, are components of a vector $\mathbf{a}$?

Chapter 5

Deformation of Continuous Media

The objective of this chapter is to present the equations that model the elastic behavior of shell-like bodies in terms of physical components as an application of the theory developed in the previous chapter. In order to do so, however, it is worthwhile to start by giving a summary of the theories of elasticity for general materials and for shells in particular. The notion of elasticity originated in the form of the law announced by Robert Hooke and succinctly explained by him in 1678 as "the power of any springy body is in the same proportion with the extension."This statement, whose precise meaning is not clear in this original form, can now be understood as the constitutive law for linear elastic materials and it was the work of great mathematical physicists of the last three centuries to develop a mathematical theory of elasticity extended to other sort of materials. The load-deflection relationship established by this law can be stated as a tensor equation relating the stress and strain, and the first two sections of this chapter are concerned with presenting this relationship for elastic bodies in general.

On the other hand, the theory of shell-like bodies has been built by engineers and mathematicians over the last hundred years or so. We shall give a precise definition below, but, colloquially, we may say that by "shell" we mean any three-dimensional solid body that is fairly thin with a curved shape (the particular case of flat-shaped bodies is referred to as plates) and that is capable of supporting and transmitting loads. A theory for the deformation of plates developed from the three-dimensional equations of linear elasticity started off in 1850 with the work of [Kirchhoff, 1850] which has become known as the Poisson–Kirchhoff theory for bending of plates. The more general theory for bending of thin shells began in a publication of 1888 due to [Love, 1888].

Since a shell is described as a three-dimensional continuum with a special boundary surface, the classical three-dimensional theory of continuum mechanics can provide a complete description of the motion and deformation of shells. However, the three-dimensional initial and boundary-valued problem in Euclidean space that model the mechanics of such bodies turns out to be computationally and analytically intractable by the mathematical tools available so far. One way around this difficulty is to content ourselves with only a partial description for sufficiently thin shells by transfering as much information as possible to a surface of reference. Since surfaces are described by parameterizations defined on subsets of a two-dimensional space, a suitable theory for shells can be constructed by means of analytically tractable two-dimensional problems so that the purpose of a theory of shells is to model shell-like bodies by finding appropriate two-dimensional equations. Historically, this has been approached over the last hundred years or so by using one of the following alternative strategies:

(1) *Derivation of two-dimensional equations from the three-dimensional theory of continuum mechanics*: This strategy consists of developing an approximate two-dimensional theory constructed in a way that can provide only partial information. On one hand, this partial information is concerned with the quantities representing the response of a deformed surface at a time t to the motion of the entire material body. On the other hand, they can be concerned with the averages of quantities resulting from the motion of the body.

(2) *Direct method*: This approach was introduced by the Cosserat brothers ([Cosserat, 1908] and [Cosserat, 1909]). It consists of representing a three-dimensional thin shell by an idealized body (its model). This is represented by a two-dimensional continuum, which is postulated a priori (that is, directly and independently of the three-dimensional equations) with the purpose to portray the thin shell. This model, called *Cosserat surface*, consists of a surface with a single deformable vector assigned to every point of the surface.

It is our aim to present the essentials of the theory of thin elastic shells as developed from the three-dimensional equations of continuum mechanics only and we do not pursue the direct method here. We should note that the summary of the theory of shells that is presented in Section 5.3 and the subsequent ones is thoroughly based on [Naghdi, 1972] and [Dikmen, 1982].

5.1 Stress tensor and equations of motion

Let a body $\mathcal{B}$ set up in a Euclidean space structure be referred to a curvilinear coordinate system $\mathbf{x} = (x^1, x^2, x^3)$ centered at a nearby point O with basis vectors $\mathbf{g} = (\mathbf{g}_1, \mathbf{g}_2, \mathbf{g}_3)$ directed along the coordinate lines. Consider an element of surface area da surrounding some point P in the body and let $d\tau$ be the infinitesimal tetrahedral volume element shown in Figure 5.1 formed by the coordinate surfaces and the surface element da.

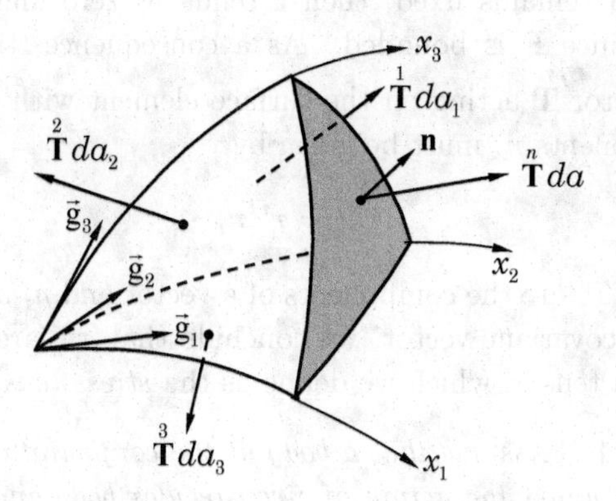

Fig. 5.1 Infinitesimal tetrahedral volume.

Taking $\mathbf{n}$ to be the unit normal to da and n_i its covariant components, the elements of area da_i in the coordinate surfaces are given by

$$da_i = n_i da. \tag{5.1}$$

We assume that the element of volume $d\tau$ is proportional to the element of surface area so that $d\tau = \ell da$, for some factor ℓ depending on the linear dimension of the volume element.

Suppose that we have a stress vector (force per unit area) $\overset{n}{\mathbf{T}} = T^j \mathbf{g}_j$ acting on an element of surface area da. Here, the superscript brings into evidence the dependence of the stress vector on the orientation $\mathbf{n}$ of the surface elements da. We assume that this force is bounded. Let $\overset{i}{\mathbf{T}}$ denote the stress vectors acting on each surface element da_i, positively oriented by the directions of the exterior normals to the volume element, and we write each of them in component form as

$$\overset{i}{\mathbf{T}} = -\tau^{ij} \mathbf{g}_j,$$

where τ^{ij} are the contravariant components of the stress vectors acting on da_i.

Since the body $\mathcal{B}$ is in a state of equilibrium under the actions of prescribed body and surface forces, the resultant of all forces and the resultant moments of these forces acting on every subregion of $\mathcal{B}$ must vanish. So, the first condition of equilibrium requires that

$$F^i \mathbf{g}_i \ell da + T^j \mathbf{g}_j da - \tau^{ij} \mathbf{g}_j da = \mathbf{0}. \tag{5.2}$$

If the point P tends to the coordinate center O in a way that the direction of the normal $\mathbf{n}$ remains fixed, then ℓ tends to zero and the first term in (5.2) vanishes since $\mathbf{F}$ is bounded. As a consequence, the components T^j of the stress vector $\overset{n}{\mathbf{T}}$ acting on the surface element with orientation $\mathbf{n}$ with covariant components n_i must be given by

$$T^j = \tau^{ij} n_i. \tag{5.3}$$

Note that, since T^i are the components of a vector and n_i are the components of an arbitrary covariant vector, we conclude that τ^{ij} are the contravariant components of a tensor, which we define as the *stress tensor*.

Proposition 5.1 *Assume that a body at the configuration shown in Figure 5.1 is in motion under the action of given regular body and surface forces per unit volume* $\mathbf{F} = F^i \mathbf{g}_i$ *acting on the mass contained in the volume element* $d\tau$ *and under the inertial forces* $-\rho_o a^i$*, with* ρ_o *being the density and* a^i *the acceleration.*

Theorem 5.1 *Then, the components of the stress tensor* τ^{ij} *at each point of the body are symmetric (i.e.,* $\tau^{ij} = \tau^{ji}$*) and satisfy the following system of partial differential equation*

$$\tau^{ij}_{\ ,\,j} + F^i = \rho_o a^i. \tag{5.4}$$

Moreover, the components T^j *of the stress vector acting on the surface element with orientation* ν *with covariant components* ν_i *must be given by (5.3),*

$$T^j = \tau^{ij} n_i.$$

By **regular** body and surface forces we mean that the components of body forces $F^i = F^i(\mathbf{x})$ are continuous functions at each point of the body and that the components T^i of the stress vector are continuously differentiable.

Clearly, for bodies in equilibrium in the case that there is no acceleration, formula (5.4) reduces to

$$\tau^{ij}_{\ ,\,j} + F^i = 0. \tag{5.5}$$

■ **Proof:** Formula (5.3) allows us to determine the stress vector acting on a surface element with the specified orientation whenever the nine functions τ^{ij} are known. Therefore, it is our purpose to derive a system of partial differential equations that will enable us to determine the components τ^{ij} of the stress tensor whenever a body is in equilibrium under the action of prescribed body and surface forces F^i.

First, consider that every portion of the body is in equilibrium and there is no acceleration. Then, both the resultant of all forces and the resultant moment of these forces acting on every subregion τ of $\mathcal{B}$ must vanish. Taking λ_i to be the unit vector in an arbitrarily fixed direction and assuming that the components of body force $F^i = F^i(\mathbf{x})$ are continuous functions at each point of the body and that the components T^i of the stress vector are continuously differentiable, the equilibrium condition yields the equation

$$\int_\tau F^i \lambda_i d\tau + \int_S T^i \lambda_i da = 0. \tag{5.6}$$

Substituting (5.3) for T^i, we have

$$\int_\tau F^i \lambda_i d\tau + \int_S \tau^{ji} n_j \; \lambda_i da = 0.$$

We can apply Gauss' divergence theorem on the second integral to yield

$$\int_\tau \left(F^i \lambda_i + \left(\tau^{ji} \lambda_i \right)_{,\, j} \right) d\tau = 0.$$

Since λ_i is a parallel vector field so that its derivatives $\lambda_{i,j}$ vanish, this equation becomes

$$\int_\tau \left(F^i + \tau^{ji}_{\; ,\, j} \right) \lambda_i \; d\tau = 0. \tag{5.7}$$

But the integrand is continuous and λ_i is arbitrary so that we necessarily have that the stress tensor τ^{ji} at each point of the body satisfies the system of partial differential equations (5.5):

$$\tau^{ji}_{\; ,\, j} + F^i = 0.$$

Now we show that the stress tensor is symmetric, that is, $\tau^{ji} = \tau^{ij}$. Indeed, the equilibrium condition also implies that the resultant moment of the body and surface forces vanishes. Thus,

$$\int_\tau \mathbf{F} \times \mathbf{r} \cdot \boldsymbol{\lambda} \; d\tau + \int_S \mathbf{T} \times \mathbf{r} \cdot \boldsymbol{\lambda} \; da = 0,$$

where $\mathbf{r} = l^i \mathbf{b}_i$ is the position vector of the point $P(\mathbf{x})$ relative to the center point O, $\mathbf{F} \times \mathbf{r} \cdot \boldsymbol{\lambda} \; d\tau$ is the component of the moment $\mathbf{F} \times \mathbf{r} \; d\tau$ in the direction of the unit vector $\boldsymbol{\lambda}$, and $\mathbf{T} \times \mathbf{r} \cdot \boldsymbol{\lambda} \; da$ is the component of the moment due to the surface forces. Applying the expression for the triple scalar product and substituting $T^i = \tau^{mi} n_m$ from (5.3), this gives

$$\int_\tau \epsilon_{ijk} F^i l^j \lambda^k \; d\tau + \int_S \epsilon_{ijk} \tau^{mi} \nu_m l^j \lambda^k \; da = 0.$$

Again, we apply Gauss' divergence theorem in the second integral to obtain

$$\int_{\tau} \epsilon_{ijk}\lambda^k \left(F^i l^j + \left(\tau^{mi} l^j\right)_{,m}\right) d\tau = 0$$

since $\epsilon_{ijk,m} = 0$. By carrying out the covariant differentiation and using (5.5) we get

$$\int_{\tau} \epsilon_{ijk}\tau^{mi} l^j_{,m}\ \lambda^k\ d\tau = 0. \tag{5.8}$$

On the other hand, since

$$\mathbf{b}_j = \frac{\partial \mathbf{r}}{\partial x^j} = l^i_{\ ,\,j}\mathbf{b}_i,$$

we have that $l^j_{,m} = \delta^j_m$. From this and the arbitrariness of the volume τ in (5.8) we must have

$$\epsilon_{ijk}\tau^{ji}\lambda^k = 0$$

or, since $\epsilon_{ijk} = -\epsilon_{jik}$, this can be expressed as

$$\frac{1}{2}\epsilon_{ijk}\left(\tau^{ji} - \tau^{ij}\right)\lambda^k = 0.$$

This can be expanded using $\epsilon_{ijk} = \sqrt{g}\ e_{ijk}$ for $\sqrt{g} \neq 0$ to yield

$$\left(\tau^{23} - \tau^{32}\right)\lambda^1 + \left(\tau^{31} - \tau^{13}\right)\lambda^2 + \left(\tau^{12} - \tau^{21}\right)\lambda^3 = 0.$$

Taking into account that the direction of $\boldsymbol{\lambda}$ is arbitrary this implies that

$$\tau^{ij} = \tau^{ji}$$

and we conclude that the stress tensor is symmetric. At this point, if we add the inertial force $-\rho_o a^i$, where ρ_o is the density and a^i is the acceleration, to the body force F^i per unit volume, the principle of D'Alembert establishes that the equations of motion are

$$\tau^{ij}_{\ ,\,j} + F^i = \rho_o a^i.$$

If we take F^i to stand for the body force per unit mass, the equations of motion become

$$\tau^{ij}_{\ ,\,j} + \rho_o F^i = \rho_o a^i. \tag{5.9}$$

All these equations are the same in all admissible reference frames because they appear in tensor form. In the reference frame of the undeformed configuration, the covariant derivatives in (5.9) should be taken with respect to the metric coefficients h_{ij} and the accelerations a^i and the forces F^i are components of the acceleration and force vectors relative to the basis in the reference configuration. □

In the following example, we derive (5.5) in spherical and cylindrical polar coordinates.

Example 5.1 Find the equations of a body in equilibrium under the action of a given force **F** in

(a) spherical polar coordinates;

(b) cylindrical polar coordinates.

Solution:

(a) Consider the spherical coordinates

$$x^1 = r, \quad x^2 = \phi, \quad x^3 = \theta,$$

and the transformation

$$x = r \sin\theta \cos\phi, \quad y = r \sin\theta \sin\phi, \quad z = r\cos\theta.$$

This transformation defines a spherical system of coordinates where the position vector is given by

$$\mathbf{r} = x\mathbf{i} + y\mathbf{j} + z\mathbf{k} = r(\sin\theta\cos\phi\ \mathbf{i} + \sin\theta\sin\phi\ \mathbf{j} + \cos\theta\ \mathbf{k}).$$

From this, we have

$$\mathbf{g}_1 = \frac{\partial \mathbf{r}}{\partial r} = \sin\theta\cos\phi\ \mathbf{i} + \sin\theta\sin\phi\ \mathbf{j} + \cos\theta\mathbf{k},$$

$$\mathbf{g}_2 = \frac{\partial \mathbf{r}}{\partial \theta} = r(\cos\theta\cos\phi\ \mathbf{i} + \cos\theta\sin\phi\ \mathbf{j} - \sin\theta\ \mathbf{k}),$$

$$\mathbf{g}_3 = \frac{\partial \mathbf{r}}{\partial \phi} = r(\sin\theta\cos\phi\ \mathbf{i} + \cos\theta\cos\phi\ \mathbf{j}),$$

and

$$\begin{aligned} g_{11} &= \mathbf{g}_1 \cdot \mathbf{g}_1 = 1, \\ g_{22} &= \mathbf{g}_2 \cdot \mathbf{g}_2 = r^2, \\ g_{33} &= \mathbf{g}_3 \cdot \mathbf{g}_3 = r^2 \sin^2\phi, \\ g_{ij} &= 0, \quad \text{if } i \neq j. \end{aligned}$$

From the last relation, we see that the coordinate system is orthogonal. Therefore, for g^{ij} we have from the definition that

$$g^{11} = 1, \quad g^{22} = \frac{1}{r^2}, \quad g^{33} = \frac{1}{r^2 \sin^2 \phi}, \quad g^{ij} = 0 \text{ if } i \neq j.$$

On the other hand, the Christoffel symbols are

$$\begin{aligned} \Gamma^1_{22} &= -r \qquad , & \Gamma^1_{33} &= -r \sin^2 \phi, \\ \Gamma^2_{12} &= \Gamma^2_{21} = \frac{1}{r}, & \Gamma^2_{33} &= -\sin\phi \cos\phi, \\ \Gamma^3_{13} &= \Gamma^3_{31} = \frac{1}{r}, & \Gamma^3_{23} &= \Gamma^3_{32} = \cot\phi, \end{aligned}$$

and

$$\Gamma^i_{jk} = 0 \quad \text{for all other } i, j, k.$$

Consider the displacement vector and let u_i be its covariant components. Then, since $\Gamma^a_{ij} = \Gamma^a_{ji}$, we have that the infinitesimal strain tensor components are

$$e_{ij} = \frac{1}{2}(u_{i;j} + u_{j;i}) = \frac{1}{2}(u_{i,\,j} + u_{j,i}) - \Gamma^a_{ij} u_a,$$

where we remind that $u_{i;j}$ denotes the anholonomic covariant derivatives

$$u_{i;j} = u_{i,j} - u_a \Gamma^a_{ij}.$$

From this, it follows that

$$\begin{aligned} e_{11} &= \frac{\partial u_1}{\partial r}, \\ e_{12} &= \frac{1}{2}\left(\frac{\partial u_1}{\partial \phi} + \frac{\partial u_2}{\partial r}\right) - \frac{1}{r} u_2, \\ e_{22} &= \frac{\partial u_2}{\partial \phi} + r u_1, \\ e_{23} &= \frac{1}{2}\left(\frac{\partial u_2}{\partial \theta} + \frac{\partial u_3}{\partial \phi}\right) - \cot\phi \; u_3, \\ e_{31} &= \frac{1}{2}\left(\frac{\partial u_1}{\partial \theta} + \frac{\partial u_3}{\partial r}\right) - \frac{1}{r} u_3, \\ e_{33} &= \frac{\partial u_3}{\partial \theta} + r \sin^2\phi \; u_1 + \sin^2\phi \cos^2\phi \; u_2. \end{aligned}$$

Let $\xi_r, \xi_\phi, \xi_\theta$ be the physical components of the displacement vector and let ε_{ij} be the physical components of the strain vector. Considering that the spherical coordinates are orthogonal, we have

$$\xi_r = \sqrt{g^{11}}\ u_1 = u_1,\ \ \xi_\phi = \sqrt{g^{22}}\ u_2 = \frac{1}{r}u_2,\ \ \xi_\theta = \sqrt{g^{33}}\ u_3 = \frac{1}{r\sin\phi}u_3,$$

and

$$\varepsilon_{ij} = \sqrt{g^{ii}g^{jj}}\ e_{ij},$$

so that the above expressions for ε_{ij} become

$$\begin{aligned}
\varepsilon_{11} &= \frac{\partial \xi_r}{\partial r},\\
\varepsilon_{12} &= \frac{1}{2}\left(\frac{1}{r}\frac{\partial \xi_r}{\partial \phi} + \frac{\partial \xi_\phi}{\partial r} - \frac{\xi_\phi}{r}\right),\\
\varepsilon_{22} &= \frac{\partial \xi_\phi}{r\partial \phi} + \frac{\xi_r}{r},\\
\varepsilon_{23} &= \frac{1}{2}\left(\frac{1}{r\sin\phi}\frac{\partial \xi_\phi}{\partial \theta} + \frac{1}{r}\frac{\partial \xi_\theta}{\partial \phi} - \frac{\cot\phi}{r}\xi_\theta\right),\\
\varepsilon_{31} &= \frac{1}{2}\left(\frac{1}{r\sin\phi}\frac{\partial \xi_r}{\partial \theta} + \frac{\partial \xi_\theta}{\partial r} - \frac{\xi_\theta}{r}\right),\\
\varepsilon_{33} &= \frac{1}{r\sin\phi}\frac{\partial \xi_\theta}{\partial \theta} + \frac{1}{r}\xi_r + \frac{\cot\phi}{r}\xi_\phi.
\end{aligned}$$

Let F_r, F_θ, F_ϕ denote the physical components of the body force vector. Then, the equations of equilibrium for the components of the stress tensor τ^{ij} become

$$\frac{1}{r}\frac{\partial}{\partial r}\left(r^2\tau^{rr}\right) + \frac{1}{r\sin\theta}\frac{\partial}{\partial \theta}\left(\sin\theta\ \tau^{r\theta}\right) + \frac{1}{r\sin\theta}\frac{\partial}{\partial \phi}\left(\tau^{r\phi}\right) - \frac{1}{r}\left(\tau^{\theta\theta} + \tau^{\phi\phi}\right) = -F_r,$$

$$\frac{1}{r^3}\frac{\partial}{\partial r}\left(r^3\tau^{r\theta}\right) + \frac{1}{r\sin^2\phi}\left(\sin^2\phi\ \tau^{\theta\phi}\right) + \frac{1}{r\sin\phi}\left(\tau^{\theta\theta}\right) = -F_\theta,$$

$$\frac{1}{r^3}\frac{\partial}{\partial r}\left(r^3\tau^{r\phi}\right) + \frac{1}{r\sin\phi}\frac{\partial}{\partial \phi}\left(\sin\phi\ \tau^{\phi\phi}\right) + \frac{1}{r\sin\phi}\frac{\partial}{\partial \theta}\left(\tau^{\theta\phi}\right) - \frac{\cot\phi}{r}\tau^{\theta\theta} = -F_\phi.$$

(b) For cylindrical polar coordinates we have

$$x^1 = r,\ \ x^2 = \theta,\ \ x^3 = z,$$

and the transformation

$$x = r\cos\theta,\ \ y = r\sin\theta,\ \ z = z.$$

From this, we have

$$g_{11} = 1, \quad g_{22} = r^2, \quad g_{33} = 1, \quad g_{ij} = 0 \text{ for all other } i, j,$$

$$g^{11} = 1, \quad g^{22} = \frac{1}{r^2}, \quad g^{33} = 1, \quad g^{ij} = 0 \text{ for all other } i, j,$$

$$\Gamma^1_{22} = -r, \quad \Gamma^2_{12} = \Gamma^2_{21} = \frac{1}{r}, \quad \Gamma^i_{jk} = 0 \text{ for all other } i, j, k.$$

Let ξ_r, ξ_θ, ξ_z denote the physical components of the displacement vector. Then, the physical components of the strain vector are

$$\varepsilon_{rr} = \frac{\partial \xi_r}{\partial r}, \quad \varepsilon_{\theta\theta} = \frac{1}{r}\frac{\partial \xi_\theta}{\partial \theta} + \frac{\xi_r}{r}, \quad \varepsilon_{zz} = \frac{\partial \xi_z}{\partial z},$$

$$2\varepsilon_{\theta z} = \frac{\partial \xi_\theta}{\partial z} + \frac{1}{r}\frac{\partial \xi_z}{\partial \theta}, \quad 2\varepsilon_{zr} = \frac{\partial \xi_z}{\partial r} + \frac{1}{r}\frac{\partial \xi_r}{\partial z}, \quad 2\varepsilon_{r\theta} = \frac{1}{r}\frac{\partial \xi_r}{\partial \theta} + \frac{\partial \xi_r}{\partial r} - \frac{\xi_\theta}{r}.$$

The equations of equilibrium for the components of the stress tensor become

$$\frac{1}{r}\frac{\partial}{\partial r}\left(r\tau^{rr}\right) + \frac{1}{r}\frac{\partial \tau^{r\theta}}{\partial \theta} + \frac{\partial \tau^{rz}}{\partial z} - \frac{\tau^{\theta\theta}}{r} = -F_r,$$

$$\frac{1}{r^2}\frac{\partial}{\partial r}\left(r^2\tau^{\theta r}\right) + \frac{1}{r}\frac{\partial \tau^{\theta\theta}}{\partial \theta} + \frac{\partial \tau^{\theta z}}{\partial z} = -F_\theta,$$

$$\frac{1}{r}\frac{\partial}{\partial r}\left(r\tau^{zr}\right) + \frac{1}{r}\frac{\partial \tau^{z\theta}}{\partial \theta} + \frac{\partial \tau^{zz}}{\partial z} = -F_z.$$

5.1.1 *Equations of motion referred to two reference frames*

In many problems of interest in continuum mechanics it is convenient to express the terms of a tensor equation with respect to a reference (undeformed) configuration rather than to a current (deformed) configuration. As an example, consider the Cauchy's first law of motion with respect to a reference configuration, expressed in its tensor form as

$$\operatorname{div} \mathbb{S} + \mathbb{P} = \rho_o \mathbf{a}, \tag{5.10}$$

and the boundary condition

$$\boldsymbol{\upsilon}\mathbb{S} = \mathbb{F}. \tag{5.11}$$

Here, $\mathbb{S}$ is the first Piola–Kirchhoff stress tensor (see Reddy [Reddy, 2013]), $\mathbb{P}$ is the referential body force, $\mathbf{a}$ is the acceleration vector, ρ_o is the mass density in the reference configuration, $\boldsymbol{\upsilon}$ is the exterior normal, and $\mathbb{F}$ is the prescribed surface traction. Since these equations are invariant under change of basis, we can represent them in component form referred to the natural basis as

$$S^{Kk}{}_{:K} + P^k = \rho_o a^k, \tag{5.12}$$

$$\upsilon_K S^{Kk} = F^k, \tag{5.13}$$

for $k = 1, 2, 3$ and $K = I, II, III$, where

$$S^{Kk}{}_{:K} = S^{Kk}{}_{,K} + S^{Km}\Gamma^k_{m\ell}x^\ell{}_{,K} + S^{Kk}\Gamma^L_{LK}. \tag{5.14}$$

In these equations, $S^{Kk}{}_{:K}$ indicates the total covariant derivative of the component S^{Kk} of the first Piola–Kirchhoff stress tensor with respect to X^K. We remind the reader that $x^\ell{}_{,K}$ are the derivatives of x^ℓ with respect to X^K and $\Gamma^k_{m\ell}$ are the Christoffel symbols. Substituting

$$\left[S^{Aa}{}_{:A}\right]_e = e^a_k S^{Kk}{}_{:K}, \quad \left[S^{Aa}\right]_{F\otimes e} = F^A_K e^a_k S^{Kk}, \tag{5.15}$$

as well as

$$[a^a]_e = e^a_k a^k, \quad [P^a]_e = e^a_k P^k,$$

$$[F^a]_e = e^a_k F^k, \quad [\upsilon_A]_{F^*} = F^K_A \upsilon_K,$$

into equations (5.12) and (5.13), we have

$$e^a_k\left(F^K_A e^k_a \left[S^{Aa}\right]_{F\otimes e}\right)_{:K} + [P^a]_e = \rho_o [a^a]_e, \tag{5.16}$$

$$[\upsilon_A]_{F^*}\left[S^{Aa}\right]_{F\otimes e} = [F^a]_e. \tag{5.17}$$

These equations are the expressions of Cauchy's first law of motion and boundary conditions expressed in terms of anholonomic components of the vectors and tensors involved. It should be noted that, for $L = K$ and $B = A$, the expression of the anholonomic components of a third-order tensor of double field, namely,

$$\left[S^{Aa}{}_{:B}\right]_{F\otimes e\otimes F^*} = F^A_K e^a_k F^L_B S^{Kk}{}_{:L}, \tag{5.18}$$

reduces to

$$\left[S^{Aa}{}_{:A}\right]_e = F^A_K e^a_k F^K_A S^{Kk}{}_{:K} = e^a_k S^{Kk}{}_{:K}, \tag{5.19}$$

which justifies the use of formulas (5.15).

The following anholonomic equations are another representation in component form:

$$\left[S^{Kk}{}_{:K}\right]_{F\otimes e\otimes F^*} + \left[P^k\right]_e = \rho_o \left[A^k\right]_e \tag{5.20}$$

and

$$[\upsilon_K]_{F^*}\left[S^{Kk}\right]_{F\otimes e} = \left[F^k\right]_e, \tag{5.21}$$

where

$$\begin{aligned}\left[S^{Kk}{}_{:K}\right]_{F\otimes e\otimes F^*} &= \left[S^{Kk}{}_{,K}\right]_{F\otimes e\otimes F^*}\\ &+ \left[S^{Km}\right]_{F\otimes e}\left[\Gamma^k_{m\ell}\right]_{e\otimes e^*\otimes e^*}\left[x^\ell{}_{,K}\right]_{e\otimes F^*}\\ &+ \left[S^{Kk}\right]_{F\otimes e}\left[\Gamma^L_{LK}\right]_{E\otimes E^*\otimes F}.\end{aligned} \tag{5.22}$$

On the other hand, the expression of the tensor components for the Green strain tensor,

$$E_{IJ} = \frac{1}{2}\left(U_{I;J} + U_{J;I} + U_{M;I}U^M{}_{;J}\right), \tag{5.23}$$

when written in terms of anholonomic components reads

$$\begin{aligned}\left[E_{IJ}\right]_{F^*\otimes E^*} = \frac{1}{2}\Big(&\left[U_{I:J}\right]_{F^*\otimes E^*} + \left[U_{J;I}\right]_{E^*\otimes F^*}\\ &+ \left[U_{M;I}\right]_{E^*\otimes F^*}\left[U^M_{;J}\right]_{E^*\otimes E^*}\Big).\end{aligned}$$

Cauchy's first law of motion can be expressed in terms of physical components by substituting the factors of basis transformation in equations (5.16) and (5.17). Thus, since

$$e^a_k = \delta^a_k\sqrt{g_{kk}}, \quad F^K_A = \delta^K_A\sqrt{G^{KK}}, \quad e^k_a = \delta^k_a\frac{1}{\sqrt{g_{kk}}},$$

we have the desired equations written in terms of physical components:

$$\delta^a_k\sqrt{g_{kk}}\left(\delta^K_A\sqrt{G^{KK}}\delta^k_a\frac{1}{\sqrt{g_{kk}}}\left[S^{Aa}\right]_{F\otimes e}\right)_{:K} + \left[P^a\right]_e = \rho_o\left[a^a\right]_e$$

and

$$[\upsilon_A]_{F^*}\left[S^{Aa}\right]_{F\otimes e} = \left[F^a\right]_e,$$

or, alternatively,

$$\sqrt{g_{aa}}\left(\frac{\sqrt{G^{AA}}}{\sqrt{g_{aa}}}\left[S^{Aa}\right]_{F\otimes e}\right)_{:A} + \left[P^a\right]_e = \rho_o\left[a^a\right]_e \tag{5.24}$$

and

$$[\upsilon_A]_{F^*}\left[S^{Aa}\right]_{F\otimes e} = \left[F^a\right]_e. \tag{5.25}$$

We recall that the adopted symbolic notation is the same for both physical and anholonomic components.

Similarly, the equations (5.20) and (5.21) can be represented in component form referred to the anholonomic basis $\mathbf{E}$ as the equations

$$\left[\hat{S}^{AC}_{;A}\right]_E + \rho_0 \left[P^C\right]_E = \rho_0 \left[\ddot{U}^C\right]_E, \tag{5.26}$$

$$\left[\nu_A\right]_{F^*} \left[\hat{S}^{AC}\right]_{F\otimes E} = \left[F^C\right]_E. \tag{5.27}$$

Since the procedure of getting these equations is the same, the above results show that the anholonomic components in equations (5.26) and (5.27) are related just in the same way as the tensor components in equations (5.12) and (5.13). Furthermore, these equations present the same formal aspect.

Referring to the basis $\mathbf{F} \otimes \mathbf{e}$, the anholonomic equations of motion are

$$\left[S^{Kk}_{\ \ :K}\right]_{F\otimes e\otimes F^*} + \left[P^k\right]_e = \rho_o \left[a^k\right]_e, \tag{5.28}$$

$$\left[\upsilon_K\right]_{F^*} \left[S^{Kk}\right]_{F\otimes e} = \left[F^k\right]_e, \tag{5.29}$$

where

$$\begin{aligned}\left[S^{Kk}_{\ \ :K}\right]_{F\otimes e\otimes F^*} &= \left[S^{Kk}_{\ \ ,K}\right]_{F\otimes e\otimes F^*} \\ &+ \left[S^{Km}\right]_{F\otimes e} \left[\Gamma^k_{m\ell}\right]_{e\otimes e^*\otimes e^*} \left[x^\ell_{\ ,K}\right]_{e\otimes F^*} \\ &+ \left[S^{Kk}\right]_{F\otimes e} \left[\Gamma^L_{LK}\right]_{E\otimes E^*\otimes F^*}.\end{aligned}$$

On the other hand, when written in terms of anholonomic components, the expression of the Green strain tensor components

$$E_{IJ} = \frac{1}{2}\left(U_{I;J} + U_{J;I} + U_{M;I}U^M_{\ ;J}\right)$$

reads

$$\begin{aligned}\left[E_{IJ}\right]_{F^*\otimes E^*} = \frac{1}{2}\Big(&\left[U_{I;J}\right]_{F^*\otimes E^*} + \left[U_{J;I}\right]_{E^*\otimes F^*} \\ &+ \left[U_{M;I}\right]_{E^*\otimes F^*} \left[U^M_{\ ;J}\right]_{E\otimes E^*}\Big),\end{aligned} \tag{5.30}$$

where

$$\left[U_{I;J}\right]_{F^*\otimes E^*} = \left[U_{I,J}\right]_{F^*\otimes E^*} - \left[U_K\right]_{F^*} \left[\Gamma^K_{IJ}\right]_{F\otimes F^*\otimes E^*}, \tag{5.31}$$

$$\left[U_{J;I}\right]_{E^*\otimes F^*} = \left[U_{J,I}\right]_{E^*\otimes F^*} - \left[U_K\right]_{F^*} \left[\Gamma^K_{JI}\right]_{F\otimes E^*\otimes F^*}, \tag{5.32}$$

$$\left[U^M_{\ ;J}\right]_{E\otimes E^*} = \left[U^M_{\ ,J}\right]_{E\otimes E^*} + \left[U^K\right]_E \left[\Gamma^M_{JK}\right]_{E\otimes E^*\otimes E^*}. \tag{5.33}$$

It is worth noting that (5.31) and (5.32) could also be written as

$$[U_{I;J}]_{F^*\otimes E^*} = [U_{I,J}]_{F^*\otimes E^*} - [U_K]_{E^*}\,[\Gamma^K_{IJ}]_{E\otimes F^*\otimes E^*},$$
$$[U_{J;I}]_{E^*\otimes F^*} = [U_{J,I}]_{E^*\otimes F^*} - [U_K]_{E^*}\,[\Gamma^K_{JI}]_{E\otimes E^*\otimes F^*},$$

but then the components $[U_K]_{E^*}$ would be referred to the basis $\mathbf{E}^*$, which is not a unit basis. In the expressions (5.30) through to (5.33), we have

$$\begin{aligned}
[E_{IJ}]_{F^*\otimes E^*} &= \frac{\sqrt{G^{II}}}{\sqrt{G_{JJ}}}E_{IJ},\\
[U_{I;J}]_{F^*\otimes E^*} &= \frac{\sqrt{G^{II}}}{\sqrt{G_{JJ}}}U_{I;J}, \qquad [U_{J;I}]_{E^*\otimes F^*} = \frac{\sqrt{G^{II}}}{\sqrt{G_{JJ}}}U_{J;I},\\
[U_K]_{F^*} &= \sqrt{G^{KK}}U_K, \qquad [U^K]_E = \sqrt{G_{KK}}U^K,\\
[\Gamma^K_{IJ}]_{F\otimes F^*\otimes E^*} &= \frac{\sqrt{G^{II}}}{\sqrt{G^{KK}}\,\sqrt{G_{JJ}}}\Gamma^K_{IJ} - \frac{\sqrt{G^{II}}}{\sqrt{G_{JJ}}}\left(\frac{1}{\sqrt{G^{JJ}}}\right)_{,I}\delta^K_J,\\
[\Gamma^K_{JI}]_{F\otimes E^*\otimes F^*} &= \frac{\sqrt{G^{II}}}{\sqrt{G^{KK}}\,\sqrt{G_{JJ}}}\Gamma^K_{JI} - \frac{\sqrt{G^{II}}}{\sqrt{G_{JJ}}}\left(\frac{1}{\sqrt{G^{II}}}\right)_{,J}\delta^K_I,\\
[\Gamma^M_{JK}]_{E\otimes E^*\otimes E^*} &= \frac{\sqrt{G_{MM}}}{\sqrt{G_{JJ}}\,\sqrt{G_{KK}}}\Gamma^M_{JK} - \frac{(\sqrt{G_{KK}})_{,J}}{\sqrt{G_{JJ}}\,\sqrt{G_{KK}}}\delta^M_K.
\end{aligned}$$

Since the covariant derivatives of vector and tensor components have also a tensor character, the above theory can be used to obtain their anholonomic counterparts. For instance, the covariant derivative of the components S^{Aa} of the first Piola–Kirchhoff stress tensor with respect to X^B, denoted by $S^{Aa}_{;B}$, reads

$$[S^{Aa}_{;B}]_{F\otimes e\otimes F^*} = \frac{\sqrt{g_{aa}}}{\sqrt{G^{AA}}}\sqrt{G^{BB}}S^{Aa}_{;B}, \tag{5.34}$$

which is the expression of the physical components of $S^{Aa}_{;B}$ referred to the basis $\mathcal{B} = F\otimes e\otimes F^*$.

5.1.2 *Equations of motion referred to convected coordinates*

Some authors prefer to use convected coordinates, which can be obtained from the natural and anholonomic coordinates by equating the numerical value of the coordinates x^k of a given point to that of X^K. For these coordinates, the following relation

$$S^{KI} = \left(\delta^I_R + U^I_{\ ;R}\right)S^{KR} \tag{5.35}$$

holds true. The symbols $U^I{}_{;R}$, S^{KI}, and S^{KR}, for $I, R, K = I, II, III$, stand for, respectively, the displacement gradient tensor components and the first and second Piola–Kirchhoff stress tensor. In this case, the component forms are given by

$$\left((\delta^I_R + U^I_{;R})S^{KR}\right)_{;R} + P^I = \rho_o \ddot{U}^I, \tag{5.36}$$

$$\upsilon_K(\delta^I_R + U^I_{;R})S^{KR} = F^I, \tag{5.37}$$

and equation (5.14) takes the form

$$S^{KI}_{;K} = S^{KI}_{,K} + S^{KM}\Gamma^I_{KM} + S^{KI}\Gamma^L_{LK}. \tag{5.38}$$

In terms of anholonomic components, equations (5.36) and (5.37) can be written as

$$E^C_I \left(E^I_C F^K_A (S^{AC} + [U^C_{;B}]_{E\otimes E^*}[S^{AB}]_{F\otimes E})\right)_{;K} + [P^C]_E = \rho_o[\ddot{U}^C]_E \tag{5.39}$$

and

$$[\upsilon_A]_{F^*}([S^{AC}]_{F\otimes E} + [U^C_{;B}]_{E\otimes E^*}[S^{AB}]_{F\otimes E}) = [F^C]_E. \tag{5.40}$$

The components are given by

$$[\ddot{U}^C]_E = E^C_I \ddot{U}^I, \quad [U^C_{;B}]_{E\otimes E^*} = E^C_I E^R_B U^I_{;R},$$
$$[S^{AC}]_{F\otimes E} = F^A_K E^C_I S^{KI}, \quad [S^{AB}]_{F\otimes E} = F^A_K E^B_R S^{KR},$$
$$[P^C]_E = E^C_I P^I, \quad [\upsilon_A]_{F^*} = F^K_A \upsilon_K, \quad [F^C]_E = E^C_I F^I.$$

At this point, if we substitute the following factors of basis transformation

$$E^C_I = \delta^C_I \sqrt{G_{II}}, \qquad E^C_I = \delta^I_C \frac{1}{\sqrt{G_{II}}}, \qquad F^K_A = \delta^K_A = \delta^K_A \sqrt{G^{KK}}$$

into equations (5.39) and (5.40), we obtain

$$\sqrt{G_{CC}}\left(\frac{\sqrt{G^{AA}}}{\sqrt{G_{CC}}}([S^{AC}]_{F\otimes E} + [U^C_{;B}]_{E\otimes E^*}[S^{AB}]_{F\otimes E})\right)_{;A}$$
$$+ [P^C]_E = \rho_o[\ddot{U}^C]_E \tag{5.41}$$

and

$$[\upsilon_A]_{F^*}([S^{AC}]_{F\otimes E} + [U^C_{;B}]_{E\otimes E^*}[S^{AB}]_{F\otimes E}) = [F^C]_E. \tag{5.42}$$

These are the equations of motion in terms of physical components and referred to convected coordinates. The components $[U^C_{;\,B}]_{E\otimes E^*}$ are given by

$$[U^C_{;B}]_{E\otimes E^*} = [U^C_{,\,B}]_{E\otimes E^*} + [U^A]_E \left[\Gamma^C_{BA}\right]_{E\otimes E^*\otimes E^*}, \tag{5.43}$$

where

$$\left[\Gamma^C_{BA}\right]_{E\otimes E^*\otimes E^*} = \frac{\sqrt{G_{CC}}}{\sqrt{G_{BB}}\,\sqrt{G_{AA}}}\Gamma^C_{BA} - \frac{1}{\sqrt{G_{BB}}\,\sqrt{G_{AA}}}\frac{\partial\left(\sqrt{G_{CC}}\right)}{\partial X^B}\delta^C_A. \tag{5.44}$$

5.2 Strain-displacement relations for elastic bodies

In the analysis of deformation of continuum media it is convenient to define deformation variables (such as strain) based on changes in length of a line segment. When a solid body undertakes deformation due to the action of forces and stresses, a point (denoted P) of the undeformed body moves to a point (denoted $\bar{\text{P}}$) in the deformed situation (see Figure 5.2). In the one-to-one correspondence of points, the state of strain is known if the change in length of each infinitesimal line ds through each point P is known and if the change in angle between any two infinitesimal lines at P is known.

We shall use a convected system,[1] where a point is labelled by coordinates $\mathbf{x} = (x^1, x^2, x^3)$ in the undeformed body. Let us take $\mathbf{R} = \mathbf{R}(x^i)$ to stand for

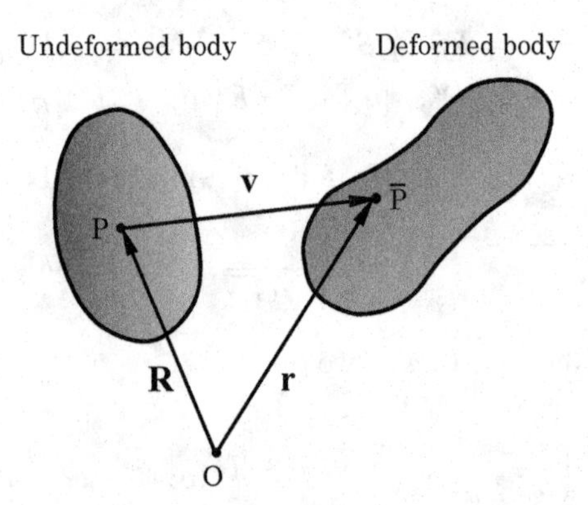

Fig. 5.2 Displacement vector **v** connecting the position vectors in the undeformed and deformed body.

[1] In the undeformed body we set up a coordinate system in a way that the coordinates x^i of a particle will remain the same even if the particle moved to a different place after deformation. Such systems are called convected systems of coordinates.

the position of P (in the undeformed body), $\mathbf{v} = \mathbf{v}(x^i)$ for the displacement of P, and

$$\mathbf{r} = \mathbf{r}(x^i) = \mathbf{R} + \mathbf{v} \tag{5.45}$$

for the position of $\bar{\text{P}}$ (in the deformed body).

As in Section 4.6, we define the strain tensor (that we denote now as γ_{ij}) and relate it to the displacement by means of equation (4.192) as

$$\gamma_{ij} = \frac{1}{2}\left(g_{ij} - G_{ij}\right), \tag{5.46}$$

where g_{ij} and G_{ij} are the metrics in the deformed and in the undeformed configurations, respectively, of a particle (point) in the body. It can be shown that this completely characterizes the state of strain in the deformed body.

We denote the components of the displacement vector field $\mathbf{v}$ relative to the bases $\mathbf{G}_i$ by V_i and the components of $\mathbf{v}$ relative to the bases $\mathbf{g}_i$ by v_i so that

$$\mathbf{v} = v_i \mathbf{g}^i, \tag{5.47}$$

$$\mathbf{v} = V_i \mathbf{G}^i. \tag{5.48}$$

As we have learned from Section 3.1.5, we can differentiate both expressions of the vector field $\mathbf{v}$ with respect to x^i to give

$$\frac{\partial \mathbf{v}}{\partial x^i} = v_{j;i}\mathbf{g}^j, \tag{5.49}$$

$$\frac{\partial \mathbf{v}}{\partial x^i} = V_{j;i}\mathbf{G}^j, \tag{5.50}$$

where

$$v_{j;i} = \frac{\partial v_j}{\partial x^i} - v_k \Gamma^k_{ji} \tag{5.51}$$

is the covariant derivative of v_j with respect to the metric g_{ij} and

$$V_{j;i} = \frac{\partial V_j}{\partial x^i} - V_k \Gamma^k_{ji} \tag{5.52}$$

is the covariant derivative of V_j with respect to the metric G_{ij}.

We can determine the metrics g_{ij} and G_{ij} by differentiating (5.45) with respect to x^i. Thus, applying the partial derivative yields

$$\mathbf{g}_i = \mathbf{G}_i + \frac{\partial \mathbf{v}}{\partial x^i}$$

so that

$$g_{ij} = \mathbf{g}_i \cdot \mathbf{g}_j = \left(\mathbf{G}_i + \frac{\partial \mathbf{v}}{\partial x^i}\right) \cdot \left(\mathbf{G}_j + \frac{\partial \mathbf{v}}{\partial x^j}\right), \tag{5.53}$$

$$G_{ij} = \mathbf{G}_i \cdot \mathbf{G}_j = \left(\mathbf{g}_i - \frac{\partial \mathbf{v}}{\partial x^i}\right) \cdot \left(\mathbf{g}_j - \frac{\partial \mathbf{v}}{\partial x^j}\right). \tag{5.54}$$

Inserting these on (5.46), we write

$$2\gamma_{ij} = \left[g_{ij} - \left(\mathbf{g}_i - \frac{\partial \mathbf{v}}{\partial x^i}\right) \cdot \left(\mathbf{g}_j - \frac{\partial \mathbf{v}}{\partial x^j}\right)\right] \tag{5.55}$$

as well as

$$2\gamma_{ij} = \left[\left(\mathbf{G}_i + \frac{\partial \mathbf{v}}{\partial x^i}\right) \cdot \left(\mathbf{G}_j + \frac{\partial \mathbf{v}}{\partial x^j}\right) - G_{ij}\right]. \tag{5.56}$$

Now, substituting in (5.50) for $\dfrac{\partial \mathbf{v}}{\partial x^i}$ in (5.53) and figuring out the dot product results in

$$2\gamma_{ij} = V_{i;j} + V_{j;i} + V^k_{;i} V_{k;j}. \tag{5.57}$$

On the other hand, substituting in (5.49) for $\dfrac{\partial \mathbf{v}}{\partial x^i}$ in (5.54) yields

$$2\gamma_{ij} = v_{i;j} + v_{j;i} + v^k_{;i} v_{k;j}. \tag{5.58}$$

Thus, these two last equations give us the strain-displacement relations in the sense that they relate γ_{ij} to the displacement $\mathbf{v}$. As one can see, they are nonlinear due to the third term on the right-hand side of each equation. We can also have the linear theory of elasticity if, for infinitesimal strains, we neglect the nonlinear terms and write

$$\gamma_{ij} = \frac{1}{2}\left(v_{j;i} + v_{i;j}\right). \tag{5.59}$$

Finally, we have some words to say about physical components. We define *the physical components of the strain* ε_{ij} as

$$\varepsilon_{ij} = \left[\gamma_{ij}\right]_{e^* \otimes e^*} = \frac{\gamma_{ij}}{\sqrt{g_{ii}}\,\sqrt{g_{jj}}}, \tag{5.60}$$

where, as we know, the γ_{ij} are the tensor components. Note that the physical components ε_{ij} are dimensionless whereas the tensor components γ_{ij} have physical dimension.

Looking up in Table 4.2, we find

$$[T_{ab}]_{e^*\otimes e^*} = \frac{T_{ab}}{\sqrt{g_{aa}}\,\sqrt{g_{bb}}},$$

where T_{ab} are the tensor components and $[T_{ab}]_{e^*\otimes e^*}$ the physical components. So, we see that the Table 4.2 gives straightaway the formula for the physical components with the bonus of giving it in a more precise way by showing which basis it is referred to. Also, for small strains, the ε_{ij} have physical meaning as follows:

$$\varepsilon_{11} \text{ is the relative elongation } E_1,$$
$$\varepsilon_{12} = \frac{\Phi_{12}}{2} \text{ is half the shearing angle.}$$

For the physical components of displacement, by normalizing all the vectors $\mathbf{g}_i$ in $\mathbf{v} = v^i\mathbf{g}_i$ we can write

$$\mathbf{v} = v^1\sqrt{g_{11}}\left(\frac{1}{\sqrt{g_{11}}}\mathbf{g}_1\right) + v^2\sqrt{g_{22}}\left(\frac{1}{\sqrt{g_{22}}}\mathbf{g}_2\right) + v^3\sqrt{g_{33}}\left(\frac{1}{\sqrt{g_{33}}}\mathbf{g}_3\right).$$

So, we define

$$u^i = \left[v^i\right]_e = \sqrt{g_{ii}}\,v^i \tag{5.61}$$

as the *physical components of the displacement* in the $\mathbf{g}_i$ directions. Note that we do not have summation in (5.61). Similarly, we can define the physical components of the displacement in the $\mathbf{g}^i$ directions as

$$u_i = \left[v_i\right]_{e^*} = \sqrt{g^{ii}}\,v_i. \tag{5.62}$$

5.3 Characterization of thin shells

We proceed to a characterization of elastic shells by giving a mathematical model while the configuration of its particles is being deformed with time. Consider a three-dimensional body $\mathcal{B}$ embedded in an open subset of the Euclidean space $\mathbb{R}^3$ and identify its particles (material points) by their positions relative to a fixed coordinate system. We shall call the *reference configuration* of $\mathcal{B}$ the configuration, at time $t = 0$, that the particles are found to be before any eventual subsequent deformation of the body takes place. Whenever some deformation occurs, if this is the case, we shall say that the particles of the body are in the *deformed configuration*.

Let (x_1, x_2, x_3) denote coordinates on a fixed rectangular Cartesian coordinate system with the positive (that is, right-handed) orientation in a Euclidean three-dimensional space and let $(\theta^1, \theta^2, \theta^3)$ denote coordinates on a general convected (that is, moving) curvilinear coordinate system related to this Cartesian system by the transformation

$$\theta^i = \theta^i(x_1, x_2, x_3). \tag{5.63}$$

Since we need this transformation to be (at least locally) invertible with inverse transformation

$$x_i = x_i(\theta^1, \theta^2, \theta^3), \tag{5.64}$$

it suffices, for this purpose, to assume that

$$\det\left[\frac{\partial x_i}{\partial \theta^j}\right] \neq 0 \tag{5.65}$$

holds in an open subset Ω of the Euclidean space.

The introduction of a convected coordinate system, although not essential, results in some simplification in the development of the study. Also, for notational convenience, it is usual to let the Latin subscripts and superscripts take the values $1, 2, 3$ and Greek subscripts and superscripts take the values $1, 2$ and to set $\theta^3 = \xi$. With this convention, (5.63) and (5.64) are rewritten as $\theta^i = \theta^i(x_i)$ and $x_i = x_i(\theta^\alpha, \xi)$, respectively. Relative to the same fixed origin (or, in other words, to the same fixed rectangular Cartesian coordinate system (x_1, x_2, x_3) introduced above), we take the functions

$$\mathbf{P} = \mathbf{P}(\theta^\alpha, \xi) \tag{5.66}$$

and

$$\mathbf{p} = \mathbf{p}(\theta^\alpha, \xi, t) \tag{5.67}$$

to represent, respectively, the position vectors of a typical particle of $\mathcal{B}$ in the reference configuration and in the deformed configuration at time t. Clearly, in view of (5.63), these vectors are equivalently expressed as functions of x_i. Thus, the t-parametric family of regions in the space that is mapped by (5.67) describes the motion (deformation) of the body $\mathcal{B}$ in such a way that these regions are the images of the function defined by (5.67), for each $t \geq 0$.

We assume that the vector function $\mathbf{p}$ is sufficiently smooth (that it has continuous partial derivatives up to any order that may be necessary for mathematical working out of the modelling). It follows that the metric tensors g_{ij}

of the coordinate system (θ^i) are given by

$$g_{ij} = \mathbf{g}_i \cdot \mathbf{g}_j \quad \text{with} \quad \mathbf{g}_i = \frac{\partial \mathbf{p}}{\partial \theta^i}. \tag{5.68}$$

From this and (5.65) we have that

$$g = \det[g_{ij}] = \left| \frac{\partial p_k}{\partial \theta^i} \cdot \frac{\partial p_k}{\partial \theta^j} \right| = \left| \frac{\partial p_k}{\partial \theta^i} \right|^2 > 0, \tag{5.69}$$

where p_k are the components of $\mathbf{p}$. By virtue of (5.69), the conjugate tensors g^{ij} (which are the inverses of g_{ij}) exist and are given by

$$g^{ij} = \mathbf{g}^i \cdot \mathbf{g}^j, \quad \text{with} \quad \mathbf{g}^i = g^{ij}\mathbf{g}_j \quad \text{and} \quad \mathbf{g}^i \cdot \mathbf{g}_j = \delta^i_j, \tag{5.70}$$

where $\mathbf{g}_i$ and $\mathbf{g}^i$ are the covariant and the contravariant basis vectors at time t, respectively. The tensors g^{ij} and g_{ij} satisfy

$$g^{ik} g_{kj} = g_{jk} g^{ki} = \delta^i_j. \tag{5.71}$$

Furthermore, consider a material surface $\mathfrak{S}$ in $\mathcal{B}$, which can be defined by the parametrization $\xi = \xi(\theta^\alpha)$ so that $\mathbf{P} = \mathbf{P}(\theta^\alpha, \xi(\theta^\alpha))$ and $\mathbf{p} = \mathbf{p}(\theta^\alpha, \xi(\theta^\alpha), t)$ represent the parametric forms of $\mathfrak{S}$ in the reference and deformed configurations, respectively. We assume that the surfaces of the one-parameter family $\mathbf{p} = \mathbf{p}(\theta^\alpha, 0, t)$ are sufficiently smooth and nonintersecting. Let this surface $\xi(\theta^\alpha) = 0$ at time t, in the deformed configuration, be referred to by s and let the position vector of a particle on this surface be denoted by $\mathbf{r}$ so that we have

$$\mathbf{r} = \mathbf{r}(\theta^\alpha, t) = \mathbf{p}(\theta^\alpha, 0, t) \tag{5.72}$$

relative to the same coordinate system to which $\mathbf{p}$ is referred. The position of any point of the shell can be specified also by giving the vector $\mathbf{r}$ and a vector $\mathbf{d}$ as shown in Figure 5.3:

$$\mathbf{p}(\theta^\alpha, \xi(\theta^\alpha), t) = \mathbf{r}(\theta^\alpha, t) + \mathbf{d}(\theta^\alpha, \xi(\theta^\alpha), t). \tag{5.73}$$

Let us denote by $\mathbf{a}_\alpha$ the basis vectors of the surface s in the present (deformed) configuration along each of the θ^α-curves, that is,

$$\mathbf{a}_\alpha = \frac{\partial \mathbf{r}}{\partial \theta^\alpha} = \mathbf{r}_{,\alpha},$$

where the comma in this notation stands for the partial differentiation with respect to the surface coordinates θ^α. Thus, we have from (5.68) that

$$\mathbf{a}_\alpha = \mathbf{g}_\alpha(\theta^\alpha, 0, t). \tag{5.74}$$

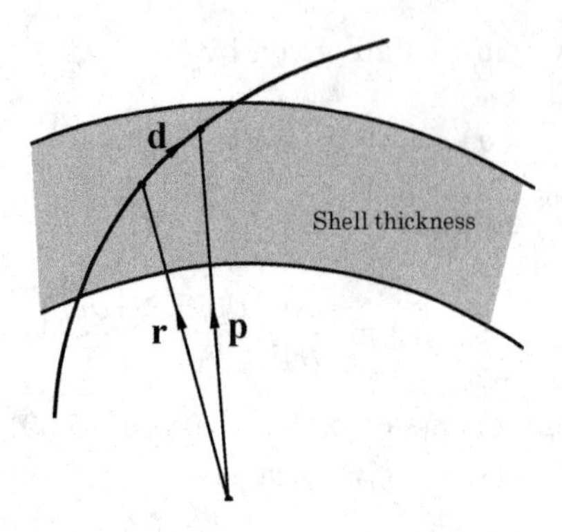

Fig. 5.3 Vectors specifying the position of any point of the shell.

We define $\mathbf{a}_3$, the unit normal to s, by the relations

$$\mathbf{a}_\alpha \cdot \mathbf{a}_3 = 0, \quad \mathbf{a}_3 \cdot \mathbf{a}_3 = 1, \quad \mathbf{a}_3 = \mathbf{a}^3, \quad a = \det(a_{\alpha\beta}) > 0, \tag{5.75}$$

where $a_{\alpha\beta} = \mathbf{a}_\alpha \cdot \mathbf{a}_\beta$ has a role in the differential geometry on the reference surface. More precisely, we can evaluate the line elements along a curve on s by means of the following computation

$$ds^2 = d\mathbf{r} \cdot d\mathbf{r} = (\mathbf{r}_{,\alpha} d\theta^\alpha) \cdot (\mathbf{r}_{,\beta} d\theta^\beta) = (\mathbf{a}_\alpha \cdot \mathbf{a}_\beta) d\theta^\alpha d\theta^\beta.$$

The quadratic form $a_{\alpha\beta} d\theta^\alpha d\theta^\beta$, where

$$\mathbf{a}_\alpha \cdot \mathbf{a}_\beta = a_{\alpha\beta}, \tag{5.76}$$

is called the *first fundamental form* of the surface. Also, we define the $a_{\alpha\beta}$ to be the covariant components of the metric tensor of s.

In differential geometry, the first fundamental form of a surface is the object which allows one to compute lengths, angles, and areas on a surface. As for a way of measuring the curvature of a surface, we need the second fundamental form. This is done by introducing the scalar product of infinitesimal vector increments

$$-d\mathbf{r} \cdot d\mathbf{a}_3 = -(\mathbf{a}_\alpha d\theta^\alpha) \cdot (\mathbf{a}_{3,\beta} d\theta^\beta) = -(\mathbf{a}_\alpha \cdot \mathbf{a}_{3,\beta}) d\theta^\alpha d\theta^\beta. \tag{5.77}$$

The quadratic form

$$b_{\alpha\beta} d\theta^\alpha d\theta^\beta,$$

where

$$b_{\alpha\beta} = -\mathbf{a}_3 \cdot \mathbf{a}_{\alpha,\beta} = \mathbf{a}_{3,\beta} \cdot \mathbf{a}_\alpha = b_{\beta\alpha}, \tag{5.78}$$

is called the *second fundamental form* of the surface s and the $b_{\alpha\beta}$ play the role of tensors in the tangent space to the surface. Note that the equality of the scalar products in (5.78) follows from the definition of the normal $\mathbf{a}_3$ to s that gives $\mathbf{a}_\alpha \cdot \mathbf{a}_3 = 0$ and

$$\mathbf{a}_{3,\beta}{\cdot}\mathbf{a}_\alpha + \mathbf{a}_3 \cdot \mathbf{a}_{\alpha,\beta} = 0.$$

Also, denoting the reciprocal base vectors of s by $\mathbf{a}^\alpha$ and putting $a^{\alpha\beta} = \mathbf{a}^\alpha \cdot \mathbf{a}^\alpha$, the geometry of Euclidean space leads to the relations

$$\mathbf{a}^\alpha = a^{\alpha\beta}\mathbf{a}_\beta, \qquad a^{\alpha\gamma}a_{\gamma\beta} = \delta^\alpha_\beta, \qquad a_{\alpha\gamma}a^{\gamma\beta}\delta^\beta_\alpha. \tag{5.79}$$

We define the *mean curvature* of the surface as the quantity

$$H = \frac{1}{2}b_{\gamma\alpha}a^{\gamma\alpha}.$$

Also,

$$K = \det\left[b_{\gamma\beta}a^{\gamma\alpha}\right] = a^{-1}\det\left[b_{\alpha\beta}\right]$$

is called the *total curvature*, or *Gaussian curvature* of the surface. Both the mean and the Gaussian curvature are surface invariants and depend only on the coefficients of the second fundamental form $b_{\alpha\beta}$. However, it can be shown that the Gaussian curvature can alternatively be expressed solely in terms of the coefficients of the first fundamental form $a_{\alpha\beta}$ and its derivatives.

Given the (local trihedron) base vectors $(\mathbf{a}_1, \mathbf{a}_2, \mathbf{a}_3)$ determined in this way at each point of the surface for which the derivatives $\mathbf{r}_{,\alpha}$ do not vanish, we can express their partial derivatives with respect to θ^α by a linear combination of themselves. Knowing that $\mathbf{a}_{3,1}$ and $\mathbf{a}_{3,2}$ are parallel to the tangent plane we can easily deduce the so-called *Weingarten formulas*:

$$\mathbf{a}_{3,\alpha} = -(b_{\kappa\alpha}a^{\kappa\gamma})\mathbf{a}_\gamma. \tag{5.80}$$

For the partial derivatives of $\mathbf{a}_1$ and $\mathbf{a}_2$, we can deduce that

$$\mathbf{a}_{\alpha,\beta} = \Gamma^\gamma_{\alpha\beta}\mathbf{a}_\gamma + b_{\alpha\beta}\mathbf{a}_3, \tag{5.81}$$

which are known as *Gauss' formulas*. Here, $\Gamma^\gamma_{\alpha\beta}$ is the Christoffel symbol given by

$$\Gamma^\gamma_{\alpha\beta} = \Gamma^\gamma_{\beta\alpha} = \mathbf{a}^\gamma{\cdot}\mathbf{a}_{\alpha,\beta} = -\mathbf{a}_\alpha{\cdot}\mathbf{a}^\gamma{}_{,\beta}. \tag{5.82}$$

Finally, we recall some other formulas from differential geometry. First, we may find convenient to use eventually a *third fundamental form*, defined by

$$d\mathbf{a}_3 \cdot d\mathbf{a}_3 = (b_{\gamma\alpha} a^{\gamma\lambda}) b_{\lambda\beta} d\theta^\alpha d\theta^\beta. \tag{5.83}$$

The components

$$(b_{\gamma\alpha} a^{\gamma\lambda}) b_{\lambda\beta} = c_{\alpha\beta} \tag{5.84}$$

can be expressed in terms of $a_{\alpha\beta}$ and $b_{\alpha\beta}$ as

$$c_{\alpha\beta} + K a_{\alpha\beta} - 2H b_{\alpha\beta} = 0. \tag{5.85}$$

Second, we have that

$$\mathbf{a}_{\alpha|\beta} = b_{\alpha\beta}\ \mathbf{a}_3, \tag{5.86}$$

$$\mathbf{a}_{3,\alpha} = -b^{\gamma}_{\alpha}\ \mathbf{a}_\gamma, \tag{5.87}$$

$$b_{\alpha\beta|\gamma} = b_{\alpha\gamma|\beta}, \tag{5.88}$$

where a vertical bar denotes covariant differentiation with respect to $a_{\alpha\beta}$.

We specify the boundary of the body $\mathcal{B}$, denoted as $\partial\mathcal{B}$, by means of the material surfaces

$$\xi = f_1(\theta^\alpha), \quad \xi = f_2(\theta^\alpha), \tag{5.89}$$

and the surface

$$F(\theta^1, \theta^2) = 0 \tag{5.90}$$

such that ξ = constant are closed smooth curves on this surface. The surface $\xi = 0$ lies between the surfaces (5.89) and, in general, f_1 and f_2 are functions of the coordinates θ^α such that $\alpha = f_1(\theta^\alpha) < 0 < \beta = f_2(\theta^\alpha)$.

The two material surfaces specified by (5.89) have the same parametric representations in all configurations because the convected coordinates $\theta^i = (\theta^\alpha, \xi)$ are identified as material coordinates. The restriction $f_1 < 0 < f_2$ for the functions in (5.89) together with (5.69) imply that the two material surfaces in (5.89) neither do not intersect each other nor the surface $\xi = 0$. In accordance with this fact and that the surface $\xi = 0$ at time t (in the deformed configuration) is referred to as s, we refer to the surfaces in the reference configuration which are part of the boundary $\partial\mathcal{B}$ and defined above by the relations (5.89) as s^- and s^+. We can choose as the reference configuration one such that the initial middle surface is midway between the surfaces given by (5.89).

The considerations so far give a mathematical characterization of a shell-like body but we can provide a less general definition of shells in a way discussed in detail in Naghdi [Naghdi, 1972] (also see Reddy [Reddy, 2004;

Reddy, 2007]). For this, let R_1 and R_2 be the principal radii of curvature of the surface s, which we assume both to be nonzero, and take $R = \min\{|R_1|, |R_2|\} > 0$. We define a neighborhood $V(s)$ of s by putting

$$V(s) = \{(\theta^\alpha, \xi) \mid \xi < R\}. \tag{5.91}$$

In this neighborhood, points in space lie along one uniquely determined normal to s. If we take the parameter ξ to denote the distance to any point in this neighborhood, measured as a function of θ^α and ξ along the positive direction of the uniquely defined normal from s, it follows that the position vector of the material points of the body $\mathcal{B}$ in the initial configuration referred to the normal coordinate system (θ^α, ξ) is given by

$$\mathbf{P} = \mathbf{P}(\theta^\alpha, \xi) = \mathbf{r}(\theta^\alpha, 0) + \xi \mathbf{a}_3(\theta^\alpha, 0), \tag{5.92}$$

where $\mathbf{r}$ and $\mathbf{a}_3$ are the position vector and the unit normal vector defined by (5.72) and (5.75), respectively.

Now, going back to the boundary $\partial\mathcal{B}$, let the bounding surfaces s^- and s^+ be specified by

$$\xi = h_1(\theta^\alpha), \quad \xi = h_2(\theta^\alpha), \quad h_1(\theta^\alpha) < 0 < h_2(\theta^\alpha), \tag{5.93}$$

$$h(\theta^\alpha) = h_2(\theta^\alpha) - h_1(\theta^\alpha), \tag{5.94}$$

where the functions h_1 and h_2 have the physical dimension of length and the surface $\xi = 0$ lies entirely between s^- and s^+.

Then, with reference to the initial configuration of the body, we define a *shell* to be a region (subset) of the space such that

$$h_1(\theta^\alpha) < \xi < h_2(\theta^\alpha), \qquad \max\{|h_1(\theta^\alpha)|, |h_2(\theta^\alpha)|\} < R,$$

bounded by the two surfaces S^- and S^+ (specified by (5.93) and called the lower and upper faces), which are situated below and above the surface $\xi = 0$ (denoted S) and an edge (or lateral) surface, specified by (5.90) in the reference configuration, such that its intersection with surfaces $\xi =$ constant is a closed smooth curve. The function $h(.)$, which is the distance between S^- and S^+ measured along $\mathbf{a}_3(\theta^\alpha, 0)$, is called the initial *thickness of the shell.* When the thickness is much smaller than the minimum radius of curvature R, that is, when

$$\frac{h}{R} \ll 1, \tag{5.95}$$

the region is called a (sufficiently) *thin shell.*

5.4 Strain-displacement relations for shells

Let us define the position vector of an arbitrary point through the thickness as $\mathbf{r} = x_i \mathbf{i}_i$, with $x_i = x_i(\xi^1, \xi^2, \xi^3 = z)$ and $\mathbf{i}_i$ being the unit vectors corresponding to the Cartesian coordinates x_i and ξ^i being the curvilinear coordinates.

First, we consider the undeformed surface. The position vector $\mathbf{R}$ can be written alternatively as

$$\mathbf{R} = \mathbf{R}_o + z\mathbf{N}, \tag{5.96}$$

where $\mathbf{R}_o$ is the position vector of a point on the undeformed middle surface and $\mathbf{N} = \mathbf{N}(\xi^1, \xi^2)$ is a unit normal vector to the undeformed middle surface. From the Vector Calculus we know that the square of the length of the line element is given by

$$(ds_o)^2 = d\mathbf{R} \cdot d\mathbf{R} = g_{ij} d\xi^i \xi^j = g_{\alpha\beta} d\xi^\alpha d\xi^\beta + g_{\alpha 3} d\xi^\alpha d\xi^3 + g_{33} d\xi^3 d\xi^3 \tag{5.97}$$

for $i, j = 1, 2, 3$ and $\alpha, \beta = 1, 2$. Since

$$d\mathbf{R} = d\mathbf{R}_o + d(z\mathbf{N}) = d\mathbf{R}_o + dz\ \mathbf{N} + z\ d\mathbf{N}$$

the square of the length becomes

$$\begin{aligned}(ds_o)^2 = {} & d\mathbf{R}_o \cdot d\mathbf{R}_o + 2dz\ \mathbf{N} \cdot d\mathbf{R}_o + 2z\ d\mathbf{N} \cdot d\mathbf{R}_o + 2z\ dz\ \mathbf{N} \cdot d\mathbf{N} \\ & + z^2 d\mathbf{N} \cdot d\mathbf{N} + (dz)^2.\end{aligned}$$

Introducing the first fundamental tensor

$$A_{\alpha\beta} = \mathbf{R}_{o,\alpha} \cdot \mathbf{R}_{o,\beta},$$

the second fundamental tensor

$$B_{\alpha\beta} = -\mathbf{N}_{,\alpha} \cdot \mathbf{a}_\beta,$$

and the third fundamental tensor

$$C_{\alpha\beta} = B^\lambda_\alpha B_{\beta\lambda},$$

we have that

$$(ds_o)^2 = (A_{\alpha\beta} - 2zB_{\alpha\beta} + z^2 C_{\alpha\beta}) d\xi^\alpha d\xi^\beta + (dz)^2, \tag{5.98}$$

since $\mathbf{N} \cdot d\mathbf{R}_o = \mathbf{N} \cdot d\mathbf{N} = 0$. From (5.97) and (5.98), we have that

$$\begin{aligned}G_{\alpha\beta} &= A_{\alpha\beta} - 2zB_{\alpha\beta} + z^2 C_{\alpha\beta}, \\ G_{\alpha 3} &= G_{3\alpha} = 0, \\ G_{33} &= 1.\end{aligned}$$

Now, we examine the situation after deformation. Considering the new position vector $\mathbf{r} = \mathbf{r}_o + \xi^3\mathbf{n} = \mathbf{r}_o + z\mathbf{n}$, the squared length of the line segment on the deformed surface is given by

$$\begin{aligned}(ds)^2 &= d\mathbf{r} \cdot d\mathbf{r} = g_{ij} d\xi^i d\xi^j \\ &= g_{\alpha\beta} d\xi^\alpha d\xi^\beta + g_{\alpha 3} d\xi^\alpha d\xi^3 + g_{33} d\xi^3 d\xi^3\end{aligned}$$

as well as by

$$\begin{aligned}(ds)^2 &= d\mathbf{r} \cdot d\mathbf{r} \\ &= d\mathbf{r}_o \cdot d\mathbf{r}_o + 2d\mathbf{r}_o \cdot \mathbf{n}\, dz + 2d\mathbf{r}_o \cdot z\, d\mathbf{n} \\ &\quad +2z\, dz\, \mathbf{n} \cdot d\mathbf{n} + (dz)^2 \mathbf{n} \cdot \mathbf{n} + z^2 d\mathbf{n} \cdot d\mathbf{n}.\end{aligned} \tag{5.99}$$

Differently from what we had before, now the vector $\mathbf{n}$ may not remain normal to the deformed surface and is no longer a unit vector so that, in principle, $\mathbf{n} \cdot d\mathbf{n} \neq 0$ and $d\mathbf{r}_o \cdot \mathbf{n} \neq 0$.

We define the tangent vector on the deformed surface as

$$\mathbf{a}_\alpha = \mathbf{r}_{o,\alpha}$$

and the following parameters

$$\begin{aligned}a_{\alpha\beta} &= \mathbf{a}_\alpha \cdot \mathbf{a}_\alpha \\ b_{\alpha\beta} &= -\mathbf{n}_{,\alpha} \cdot \mathbf{a}_\beta \\ c_{\alpha\beta} &= \mathbf{n}_{,\alpha} \cdot \mathbf{n}_{,\beta} = b_\alpha^\lambda b_{\beta\lambda}\end{aligned}$$

which we call the metric tensor (first fundamental tensor), the curvature tensor (second fundamental tensor) and the third fundamental tensor on the deformed surface, respectively. Hence, we have from (5.99) that

$$\begin{aligned}(ds)^2 &= (a_{\alpha\beta} - 2zb_{\alpha\beta} + z^2 c_{\alpha\beta}) d\xi^\alpha d\xi^\beta \\ &\quad +2(\mathbf{a}_\alpha + z\mathbf{n}_{,\alpha}) \cdot \mathbf{n}\, d\xi^\alpha dz + (dz)^2 \mathbf{n} \cdot \mathbf{n}\end{aligned} \tag{5.100}$$

We denote

$$\begin{aligned}g_{\alpha\beta} &= a_{\alpha\beta} - 2zb_{\alpha\beta} + z^2 c_{\alpha\beta} \\ g_{\alpha 3} &= 2(\mathbf{a}_\alpha + z\mathbf{n}_{,\alpha}) \cdot \mathbf{n} \\ g_{33} &= \mathbf{n} \cdot \mathbf{n}.\end{aligned}$$

Love introduced the following simplifying assumptions applicable for thin shells, which were previously introduced for plates by Kirchhoff and for beams by Bernoulli and Euler (see Reddy [Reddy, 2004; Reddy, 2007]):

(1) Points on the normal to the undeformed surface remain on the normal to the surface after the deformation takes place.
(2) These points do not change their distance from the middle surface.
(3) The normal stress on the plane parallel to the middle surface may be neglected in comparison to other stresses.

We refer to these simplifying assumptions as *Kirchhoff assumptions*. The assumption that the unit vector normal to the undeformed surface remains unitary and normal to the deformed surface, yielding $G_{\alpha 3} = 0$ and $G_{33} = 1$. Without this hypothesis, we can write the difference between the squared lengths of the line segments on the deformed and on the undeformed surfaces as

$$\begin{aligned}(ds)^2 - (ds_o)^2 = {} & [(A_{\alpha\beta} - a_{\alpha\beta}) - 2z(B_{\alpha\beta} - b_{\alpha\beta}) \\ & + z^2(C_{\alpha\beta} - c_{\alpha\beta})]d\xi^\alpha d\xi^\beta \\ & + 2(\mathbf{N} \cdot \mathbf{A}_\alpha + z\mathbf{N} \cdot \mathbf{N}_{,\alpha})d\xi^\alpha dz \\ & + (\mathbf{N} \cdot \mathbf{N})(dz)^2 - (dz)^2.\end{aligned} \tag{5.101}$$

We can define the components of the strain tensor $\gamma_{ij} = (\gamma_{\alpha\beta}, \gamma_{\alpha 3}, \gamma_{33})$ by putting

$$\gamma_{\alpha\beta} = \frac{1}{2}\left[(A_{\alpha\beta} - a_{\alpha\beta}) - 2z(B_{\alpha\beta} - b_{\alpha\beta}) + z^2(C_{\alpha\beta} - c_{\alpha\beta})\right], \tag{5.102}$$

$$\gamma_{\alpha 3} = \mathbf{N} \cdot \mathbf{A}_\alpha + z\mathbf{N} \cdot \mathbf{N}_{,\alpha}, \tag{5.103}$$

$$\gamma_{33} = \frac{1}{2}(\mathbf{N} \cdot \mathbf{N} - 1). \tag{5.104}$$

Then (5.101) becomes

$$\begin{aligned}(ds)^2 - (ds_o)^2 &= 2\gamma_{ij} d\xi^i \xi^j \\ &= 2\gamma_{\alpha\beta} d\xi^\alpha d\xi^\beta + 2\gamma_{\alpha 3} d\xi^\alpha dz + 2\gamma_{33}(dz)^2.\end{aligned}$$

Define $e_{\alpha\beta}, \chi_{\alpha\beta}, \Lambda_{\alpha\beta}$, the components of the membrane, bending, and higher-order bending strain tensors, respectively, by setting

$$\begin{aligned}e_{\alpha\beta} &= \frac{1}{2}(A_{\alpha\beta} - a_{\alpha\beta}), \\ \chi_{\alpha\beta} &= -(B_{\alpha\beta} - b_{\alpha\beta}), \\ \Lambda_{\alpha\beta} &= \frac{1}{2}(C_{\alpha\beta} - c_{\alpha\beta}).\end{aligned}$$

Thus, (5.102) can be rewritten as

$$\gamma_{\alpha\beta} = e_{\alpha\beta} + z\chi_{\alpha\beta} + z^2\Lambda_{\alpha\beta}. \tag{5.105}$$

If we consider a point Q on the undeformed middle surface and take a point P on the normal $\mathbf{N}$ at Q to this undeformed middle surface, then (see Figure 5.4)

$$\mathbf{P} = \mathbf{Q} + \zeta\mathbf{N},$$

where ζ is the distance from Q to P. Similarly, in the deformed shell, we consider a point $\bar{\text{Q}}$ on the middle surface and take a point $\bar{\text{P}}$ on the normal $\mathbf{n}$ at $\bar{\text{Q}}$ to the deformed middle surface, distance ζ from $\bar{\text{Q}}$ to $\bar{\text{P}}$, so that

$$\mathbf{p} = \mathbf{q} + \zeta\mathbf{n}.$$

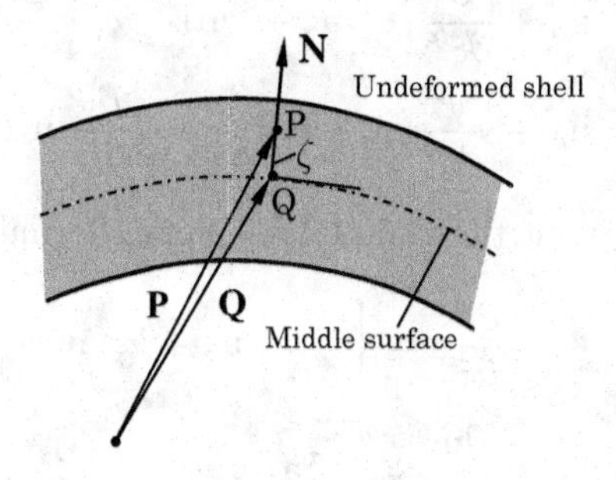

Fig. 5.4 The position vectors to a point on the undeformed middle surface and to a point on the normal.

Let $\mathbf{v} = \mathbf{v}(\xi^1, \xi^2, \zeta)$ denote the oriented segment from point $\bar{\text{P}}$ to P and let $\mathbf{v}^o = \mathbf{v}^o(\xi^1, \xi^2)$ stand for the displacement of the middle surface (the oriented segment from $\bar{\text{Q}}$ to Q). Then,

$$\mathbf{v}(\xi^1, \xi^2, \zeta) = \mathbf{v}^o(\xi^1, \xi^2) + \zeta(\mathbf{N} - \mathbf{n}). \tag{5.106}$$

Bringing on the basis vectors $(\mathbf{a}_1, \mathbf{a}_2, \mathbf{n})$ for the coordinate system centered at $\bar{\text{Q}}$ as well as the basis vectors $(\mathbf{A}_1, \mathbf{A}_2, \mathbf{N})$, centered at Q, we can write

$$\mathbf{v}^o = v^{o\gamma}\mathbf{a}_\gamma + w\mathbf{n}, \tag{5.107}$$

where w is the coordinate of $\mathbf{v}^o$ on the direction of the normal. Then, it follows that, with reference to a fixed Cartesian system of coordinates, the position vectors $\mathbf{r}_o$ and $\mathbf{R}_o$ of $\bar{\text{Q}}$ and Q, respectively, satisfy the relation

$$\mathbf{R}_o(\xi^1, \xi^2) = \mathbf{r}_o(\xi^1, \xi^2) + \mathbf{v}^o(\xi^1, \xi^2). \tag{5.108}$$

Example 5.2 Maintaining the Kirchhoff assumptions, obtain $\mathbf{A}_\alpha$ and the exact nonlinear expression of the normal $\mathbf{N}$ in the undeformed configuration in terms of the basis vectors.

Solution: The basis vectors are given by

$$\begin{aligned}\mathbf{A}_\alpha &= \frac{\partial \mathbf{R}_o}{\partial \xi^\alpha} \\ &= \frac{\partial \mathbf{r}_o}{\partial \xi^\alpha} + \frac{\partial \mathbf{v}^o}{\partial \xi^\alpha} \\ &= \mathbf{a}_\alpha + \frac{\partial \mathbf{v}^o}{\partial \xi^\alpha} \\ &= \mathbf{a}_\alpha + \frac{\partial}{\partial \xi^\alpha}\left(v^{o\gamma}\mathbf{a}_\gamma + w\mathbf{n}\right) \\ &= \mathbf{a}_\alpha + \frac{\partial v^{o\gamma}}{\partial \xi^\alpha}\mathbf{a}_\gamma + v^{o\gamma}\frac{\partial \mathbf{a}_\gamma}{\partial \xi^\alpha} + \frac{\partial w}{\partial \xi^\alpha}\mathbf{n} + w\frac{\partial \mathbf{n}}{\partial \xi^\alpha}\,. \end{aligned} \tag{5.109}$$

Introducing at this point the Gauss–Weingarten formulas

$$\frac{\partial \mathbf{a}_\gamma}{\partial \xi^\alpha} = \left\{ {\sigma \atop \gamma\,\alpha} \right\} \mathbf{a}_\sigma + b_{\gamma\alpha}\mathbf{n}, \tag{5.110}$$

$$\frac{\partial \mathbf{n}}{\partial \xi^\alpha} = -b^\sigma_\alpha \mathbf{a}_\sigma, \tag{5.111}$$

where $\left\{ {\sigma \atop \gamma\;\alpha} \right\} = \mathbf{a}^\sigma \cdot \frac{\partial \mathbf{a}_\gamma}{\partial \xi^\alpha}$, and substituting in (5.109) give

$$\mathbf{A}_\alpha = \mathbf{a}_\alpha + \frac{\partial v^{o\gamma}}{\partial \xi^\alpha}\mathbf{a}_\gamma + v^{o\gamma}\left\{ {\sigma \atop \gamma\;\alpha} \right\}\mathbf{a}_\sigma + v^{o\gamma} b_{\gamma\alpha}\mathbf{n} + \frac{\partial w}{\partial \xi^\alpha}\mathbf{n} - w b^\sigma_\alpha \mathbf{a}_\sigma.$$

Since

$$v^{o\gamma}\left\{ {\sigma \atop \gamma\;\alpha} \right\} = v^{o\sigma}{}_{,\alpha},$$

we have

$$\mathbf{A}_\alpha = \mathbf{a}_\alpha + \left(v^{o\sigma}{}_{,\alpha} - w b^\sigma_\alpha\right)\mathbf{a}_\alpha + \left(\frac{\partial w}{\partial \xi^\alpha} + v^{o\gamma} b_{\gamma\alpha}\right)\mathbf{n}. \tag{5.112}$$

Denoting

$$v^{o\sigma}{}_{,\alpha} - w b^\sigma_\alpha = \psi^\sigma_\alpha, \tag{5.113}$$

$$\frac{\partial w}{\partial \xi^\alpha} + v^{o\gamma} b_{\gamma\alpha} = \theta_\alpha, \tag{5.114}$$

the expression (5.112) for $\mathbf{A}_\alpha$ is

$$\mathbf{A}_\alpha = \mathbf{a}_\alpha + \psi^\sigma_\alpha \mathbf{a}_\alpha + \theta_\alpha \mathbf{n}. \tag{5.115}$$

As for the normal vector $\mathbf{N}$, we know that

$$\mathbf{N} = \frac{\mathbf{A}_1 \times \mathbf{A}_2}{\sqrt{A}}$$

so that

$$\mathbf{N} = \frac{1}{\sqrt{A}} \left((1 + \psi^1_1)\mathbf{a}_1 + \psi^2_1 \mathbf{a}_2 + \theta_1 \mathbf{n} \right) \times \left(\psi^1_2 \mathbf{a}_1 + (1 + \psi^2_2)\mathbf{a}_2 + \theta_2 \mathbf{n} \right).$$

On the other hand, recall that

$$\mathbf{a}_1 \times \mathbf{a}_2 = \sqrt{a}\, \mathbf{n}, \quad \mathbf{a}_2 \times \mathbf{n} = \sqrt{a}\, \mathbf{a}^1, \quad \mathbf{n} \times \mathbf{a}_1 = \sqrt{a}\, \mathbf{a}^2.$$

Then,

$$\begin{aligned} \mathbf{N} = \sqrt{\frac{a}{A}} \Big\{ & (1 + \psi^1_1)(1 + \psi^2_2)\, \mathbf{n} - (1 + \psi^1_1)\, \theta_2 \mathbf{a}^2 \\ & - \psi^2_1 \psi^1_2 \mathbf{n} + \psi^2_1 \theta_2 \mathbf{a}^1 + \theta_1 \psi^1_2 \mathbf{a}^2 - \theta_1 (1 + \psi^2_2) \mathbf{a}^1 \Big\}, \end{aligned}$$

which gives

$$\begin{aligned} \mathbf{N} = \sqrt{\frac{a}{A}} \Big\{ & (1 + \psi^1_1 + \psi^2_2 + \psi^1_1 \psi^2_2 - \psi^1_2 \psi^2_1)\mathbf{n} \\ & + \left(-\theta_1 - \theta_1 \psi^2_2 + \theta_2 \psi^2_1 \right) \mathbf{a}^1 + \left(-\theta_2 - \theta_2 \psi^1_1 + \theta_1 \psi^1_2 \right) \mathbf{a}^2 \Big\}. \end{aligned} \tag{5.116}$$

5.5 Kinematic relations for shells

In order to derive the kinematics of a shell from the three-dimensional theory for deformable continua, we assume that the position vector $\mathbf{p}(\theta^\alpha, \xi, t)$ of a typical particle of the shell at time t is an analytic function of the variable ξ on the region $h_1(\theta^\alpha) < \xi < h_2(\theta^\alpha)$ and, as such, is represented by a Taylor series expanded around $\xi_o = 0$,

$$\mathbf{p}(\theta^\alpha, \xi, t) = \mathbf{r}(\theta^\alpha, t) + \sum_{N=1}^{\infty} \xi^N \mathbf{d}_N(\theta^\alpha, t). \tag{5.117}$$

Here, (θ^α, ξ) represents a system of convected normal coordinates as introduced above, $\mathbf{r}(\theta^\alpha, t) = \mathbf{p}(\theta^\alpha, 0, t)$ is the position vector of the surface $\xi = 0$ in the deformed configuration, and the three-dimensional vector fields $\mathbf{d}_N(\theta^\alpha, t)$ are the Taylor coefficients. We take $\mathbf{d}_N$ and their partial derivatives with respect to θ^α to be referred to the basis vectors $\mathbf{a}_i$. The position vector in a reference undeformed configuration, that is, at $t = 0$, is represented by

$$\mathbf{P}(\theta^\alpha, \xi) = \mathbf{R}(\theta^\alpha) + \sum_{N=1}^{\infty} \xi^N \mathbf{D}_N(\theta^\alpha), \tag{5.118}$$

where $\mathbf{R}(\theta^\alpha) = \mathbf{r}(\theta^\alpha, 0)$ is the position vector of the surface $\xi = 0$ in the reference configuration and $\mathbf{D}_N(\theta^\alpha) = \mathbf{d}_N(\theta^\alpha, 0)$.

To account for physically admissible motions, we assume that

$$\det\left[\frac{\partial \mathbf{p}}{\partial \theta^i} \cdot \frac{\partial \mathbf{p}}{\partial \theta^j}\right] > 0. \tag{5.119}$$

Also, we assume that the two series may be differentiated with respect to all their variables as many times as required in the open region

$$h_1(\theta^\alpha) < \xi < h_2(\theta^\alpha).$$

The formulas involving the basis vectors $\mathbf{a}_\alpha$ and $\mathbf{a}_3$ as well as their reciprocal and the first and second fundamental forms developed in the above characterization of shells remain valid and depend on t. We use capital Latin letters to designate the corresponding concepts in the reference configuration (i.e., initial, $t = 0$, undeformed configuration). In particular, its first and second fundamental forms are designated by $A_{\alpha\beta}$ and $B_{\alpha\beta}$, respectively.

Let the vector field

$$\mathbf{v}^* = \frac{\partial \mathbf{p}(\theta^\alpha, \xi, t)}{\partial t} \triangleq \dot{\mathbf{p}}(\theta^\alpha, \xi, t) \tag{5.120}$$

represent the velocity vector of the three-dimensional continuum at time t. Thus, we have

$$\mathbf{v}^* = \mathbf{v} + \sum_{N=1}^{\infty} \xi^N \boldsymbol{\omega}_N, \tag{5.121}$$

with $\mathbf{v} = \dot{\mathbf{r}}$ and $\boldsymbol{\omega}_N = \dot{\mathbf{d}}_N$.

When referred to the vectors $\mathbf{a}_i$ of the basis $(\mathbf{a}_\alpha, \mathbf{a}_3)$ of s, which is the surface $\xi = 0$ in the deformed configuration, the velocity may be written as

$$\mathbf{v}^* = v^i \mathbf{a}_i = v_i \mathbf{a}^i. \tag{5.122}$$

It is convenient to introduce the notation

$$\mathbf{d} = \mathbf{d}_1, \quad \mathbf{D} = \mathbf{D}_1, \quad \boldsymbol{\omega} = \boldsymbol{\omega}_1, \tag{5.123}$$

and some new kinematical variables. Referring $\mathbf{d}_N$ and their partial derivatives with respect to θ^α to the basis vectors $\mathbf{a}_i$, one can show that

$$\begin{aligned} &\mathbf{d}_N = d_{Ni}\,\mathbf{a}^i = d_{N^{\cdot}_{\cdot}}{}^{i}\,\mathbf{a}_i, \quad d_{N}{}^{\gamma}_{\cdot} = a^{\gamma\beta} d_{N\beta}, \\ &d_{N}{}^{3}_{\cdot} = d_{N3}, \quad \frac{\partial \mathbf{d}_N}{\partial \theta^\alpha} = \lambda_{Ni\alpha}\mathbf{a}^i \quad (N \geq 2), \\ &\mathbf{d}_1 = \mathbf{d} = d_i\mathbf{a}^i = d^i\mathbf{a}_i, \quad d^\gamma = a^{\gamma\beta} d_\beta, \\ &d^3 = d_3, \quad \frac{\partial \mathbf{d}}{\partial \theta^\alpha} = \lambda_{1i\alpha}\mathbf{a}^i = \lambda_{i\alpha}\mathbf{a}^i, \end{aligned} \tag{5.124}$$

where the components $\lambda_{Ni\alpha}$ are given by

$$\begin{aligned} &\lambda_{N\gamma\alpha} = \mathbf{a}_\gamma \cdot \mathbf{d}_{N,\alpha} = d_{N_{\gamma|\alpha}} - b_{\gamma\alpha} d_{N3}, \qquad \lambda_{N}{}^{\gamma}_{\cdot\alpha} = a^{\gamma\beta}\lambda_{N\beta\alpha}, \\ &\lambda_{N3\alpha} = \mathbf{a}_3 \cdot \mathbf{d}_{N,\alpha} = d_{N_{3,\alpha}} + b^\gamma_\alpha d_{N\gamma}, \qquad \lambda_{N}{}^{3}_{\cdot\alpha} = \lambda_{N3\alpha}, \end{aligned} \tag{5.125}$$

and, for $N = 1$, given by

$$\begin{aligned} &\lambda_{\gamma\alpha} = \lambda_{|\gamma\alpha} = d_{\gamma|\alpha} - b_{\gamma\alpha} d_3, \quad \lambda^{\gamma}_{\cdot\alpha} = a^{\gamma\beta}\lambda_{\beta\alpha}, \\ &\lambda_{3\alpha} = \lambda_{|3\alpha} = d_{3,\alpha} + b^\gamma_\alpha d_\gamma, \quad \lambda^{3}_{\cdot\alpha} = \lambda_{3\alpha}. \end{aligned} \tag{5.126}$$

Let us introduce the following variables:

$$2e_{\alpha\beta} = a_{\alpha\beta} - A_{\alpha\beta} \tag{5.127}$$

$$\begin{aligned} &\varkappa_{Ni\alpha} = \lambda_{Ni\alpha} - \Lambda_{Ni\alpha}, \qquad \gamma_{Ni} = d_{Ni} - D_{Ni}, \\ &\varkappa_{i\alpha} = \varkappa_{1i\alpha} = \lambda_{i\alpha} - \Lambda_{i\alpha}, \qquad \gamma_i = \gamma_{1i} = d_i - D_i, \end{aligned} \tag{5.128}$$

where D_{Ni} and $\Lambda_{Ni\alpha}$, the corresponding values in the reference configuration to d_{Ni} and $\lambda_{Ni\alpha}$, are given by

$$\begin{aligned} &D_{Ni} = \mathbf{A}_i \cdot \mathbf{D}_N, \qquad D_i = D_{1i} = \mathbf{A}_i \cdot \mathbf{D}_1, \\ &\Lambda_{Ni\alpha} = \mathbf{A}_i \cdot \frac{\partial \mathbf{D}_N}{\partial \theta^\alpha}, \quad \Lambda_{i\alpha} = \Lambda_{1i\alpha} = \mathbf{A}_i \cdot \frac{\partial \mathbf{D}_1}{\partial \theta^\alpha}, \end{aligned} \tag{5.129}$$

where $\mathbf{A}_i$ are the vectors of the basis $(\mathbf{A}_\gamma, \mathbf{A}_3)$ and the unit normal of the surface S (the surface $\xi = 0$ in the reference configuration). We should observe that one can easily check that, under superposed rigid body motions, the quantities d_{Ni} and $\lambda_{Ni\alpha}$ remain unaltered.

Now we move on to obtain the components of a three-dimensional strain measure in terms of the variables (5.124) referred to the basis vectors $\mathbf{a}^i$. Since the basis vectors $\mathbf{g}_i$ can be written as

$$\mathbf{g}_\alpha = \mathbf{a}_\alpha + \sum_{N=1}^{\infty} \xi^N \frac{\partial \mathbf{d}_N}{\partial \theta^\alpha}, \qquad \mathbf{g}_3 = \sum_{N=1}^{\infty} N\xi^{N-1}\mathbf{d}_N, \tag{5.130}$$

with $\mathbf{a}_\alpha$ being the basis vectors defined by (5.74) of the surface that we get when $\xi = 0$ in the deformed configuration, we express the components g_{ij} as follows:

(1) Expression for $g_{\alpha\beta}$:

$$\begin{aligned} g_{\alpha\beta} &= \left(\mathbf{a}_\alpha + \sum_{N=1}^{\infty} \xi^N \frac{\partial \mathbf{d}_N}{\partial \theta^\alpha}\right)\left(\mathbf{a}_\beta + \sum_{M=1}^{\infty} \xi^M \frac{\partial \mathbf{d}_M}{\partial \theta^\beta}\right) \\ &= a_{\alpha\beta} + \xi\left(\mathbf{a}_\beta \cdot \frac{\partial \mathbf{d}_1}{\partial \theta^\alpha} + \mathbf{a}_\alpha \cdot \frac{\partial \mathbf{d}_1}{\partial \theta^\beta}\right) \\ &+ \sum_{P=2}^{\infty} \xi^P \left[\mathbf{a}_\beta \cdot \frac{\partial \mathbf{d}_P}{\partial \theta^\alpha} + \mathbf{a}_\alpha \cdot \frac{\partial \mathbf{d}_P}{\partial \theta^\beta} + \sum_{M=1}^{P-1}\left(\frac{\partial \mathbf{d}_{P-M}}{\partial \theta^\alpha} \cdot \frac{\partial \mathbf{d}_M}{\partial \theta^\beta}\right)\right] \end{aligned} \tag{5.131}$$

(2) Expression for $g_{\alpha 3}$:

$$\begin{aligned} g_{\alpha 3} &= \left(\mathbf{a}_\alpha + \sum_{N=1}^{\infty} \xi^N \frac{\partial \mathbf{d}_N}{\partial \theta^\alpha}\right)\left(\sum_{M=1}^{\infty} M\xi^{M-1}\mathbf{d}_M\right) \\ &= \mathbf{a}_\alpha \cdot \mathbf{d}_1 + \sum_{P=2}^{\infty} \xi^{P-1}\left[P\mathbf{a}_\alpha \cdot \mathbf{d}_P + \sum_{M=1}^{P-1} M\frac{\partial \mathbf{d}_{P-M}}{\partial \theta^\alpha} \cdot \mathbf{d}_M\right]. \end{aligned} \tag{5.132}$$

(3) Expression for g_{33}:

$$\begin{aligned} g_{33} &= \left(\sum_{N=1}^{\infty} N\xi^{N-1}\mathbf{d}_N\right)\left(\sum_{M=1}^{\infty} M\xi^{M-1}\mathbf{d}_M\right) \\ &= \sum_{P=2}^{\infty} \xi^{P-2}\left(\sum_{M=1}^{P-1}(P-M)\mathbf{d}_{P-M} \cdot \mathbf{d}_M\right). \end{aligned} \tag{5.133}$$

Similar results for the components G_{ij} can be obtained with the help of (5.118). Fields and functions which are defined on a three-dimensional body are distinguished from the corresponding ones defined on the two-dimensional manifold by adding an asterisk to the symbol representing the field and function on the three-dimensional body.

The covariant components of a strain measure in the three-dimensional nonlinear theory, namely the numbers γ^*_{ij} given by

$$\gamma^*_{ij} = \frac{1}{2}(\mathbf{g}_i \cdot \mathbf{g}_j - \mathbf{G}_i \cdot \mathbf{G}_j) = \frac{1}{2}(g_{ij} - G_{ij}), \tag{5.134}$$

can now be expressed as depending on the components of $\mathbf{d}_N$ and $\dfrac{\partial \mathbf{d}_N}{\partial \theta^\alpha}$ as follows:

$$
\begin{aligned}
2\gamma^*_{\alpha\beta} &= 2e_{\alpha\beta} + \xi(\varkappa_{\beta\alpha} + \varkappa_{\alpha\beta}) + \sum_{P=2}^{\infty} \xi^P \Big\{ \varkappa_{P\beta\alpha} + \varkappa_{P\alpha\beta} \\
&\quad + \sum_{M=1}^{P-1} \left[\mathbf{d}_{(P-M),\alpha} \cdot \mathbf{d}_{M,\beta} - \mathbf{D}_{(P-M),\alpha} \cdot \mathbf{D}_{M,\beta} \right] \Big\}, \\
2\gamma^*_{\alpha 3} &= \gamma_\alpha + \sum_{P=2}^{\infty} \xi^{P-1} \Big\{ P\gamma_{P\alpha} \\
&\quad + \sum_{M=1}^{P-1} M \left[\mathbf{d}_{(P-M),\alpha} \cdot \mathbf{d}_M - \mathbf{D}_{(P-M),\alpha} \cdot \mathbf{D}_M \right] \Big\}, \\
2\gamma^*_{33} &= \sum_{P=2}^{\infty} \xi^{P-2} \left\{ \sum_{M=1}^{P-1} (P-M)M \left(\mathbf{d}_{(P-M)} \cdot \mathbf{d}_M - \mathbf{D}_{(P-M)} \cdot \mathbf{D}_M \right) \right\}.
\end{aligned} \tag{5.135}
$$

Some additional results valid in the reference configuration can be found by noting that the convected coordinates θ^i can always be chosen in the reference configuration in a way such that $\mathbf{D}_N = 0$ for $N \geq 2$. With this choice, there is no loss of generality in writing the position vector in the reference configuration of $\mathcal{B}$ as

$$\mathbf{P} = \mathbf{R}(\theta^\alpha) + \xi \mathbf{D}(\theta^\alpha), \tag{5.136}$$

with $\mathbf{D}$ specified by

$$D_\alpha = 0, \quad D_3 = D, \quad \mathbf{D} = D\mathbf{A}_3. \tag{5.137}$$

The basis vectors and the metric tensor in the initial reference configuration are

$$
\begin{aligned}
\mathbf{G}_\alpha &= \mathbf{A}_\alpha + \xi \mathbf{D}_{,\alpha} = \nu^\gamma_\alpha \mathbf{A}_\gamma + \xi D_{,\alpha} \mathbf{A}_3, \\
\mathbf{G}_3 &= D\mathbf{A}_3, \\
G_{\alpha\beta} &= \nu^\gamma_\alpha \nu^\delta_\beta A_{\gamma\delta} + \xi^2 D_{,\alpha} D_{,\beta}, \\
G_{\alpha 3} &= \xi D\, D_{,\alpha} = \frac{1}{2} \xi (D^2)_{,\alpha}, \\
G_{33} &= D^2.
\end{aligned}
$$

Here, ν^γ_α is given by

$$\nu^\gamma_\alpha = \delta^\gamma_\alpha - \xi D B^\gamma_\alpha. \tag{5.138}$$

In view of this procedure, the functions $\Lambda_{Ni\alpha}$ in (5.129) reduce to

$$\Lambda_{\beta\alpha} = \Lambda_{1\beta\alpha} = -DB_{\alpha\beta}, \quad \Lambda_{3\alpha} = \Lambda_{13\alpha} = D_{,\alpha}, \tag{5.139}$$

and

$$\Lambda_{Ni\alpha} = 0, \quad \text{for } N \geq 2. \tag{5.140}$$

5.5.1 *Linearized kinematics*

In order to simplify substantially the calculations, we consider the case when $D = 1$ and proceed further towards a linearized version of the kinematical results. Working up from (5.117), it turns out that the kinematic variables occur as independent of ξ. Therefore, they can be referred to as two-dimensional variables so that the resulting two-dimensional quantities may be regarded as an exact characterization of the kinematic of shells. However, this characterization has the drawback of forming an undesirable infinite set of variables and, therefore, suitable approximations are required in order to yield useful measures of deformation of shells. This task is carried out by means of first proceeding to a linearization which is then followed by an approximation scheme.

At this point, we remind the reader that we are specifying the position vector in the reference configuration of $\mathcal{B}$ by (5.136). Thus, the values for (5.129) in the undeformed configuration become

$$D_{Ni} = 0 \quad \text{and} \quad \Lambda_{Ni\alpha} = 0, \text{ for } N \geq 2. \tag{5.141}$$

Assuming $D = 1$ for simplification purposes, we have

$$\mathbf{D} = \mathbf{A}_3, \;\; D_\alpha = 0, \;\; \Lambda_{\beta\alpha} = -B_{\alpha\beta}, \;\; \Lambda_{3\alpha} = 0. \tag{5.142}$$

This simplification gives that the linearized two-dimensional kinematic measures resulting from (5.127) and (5.128) are[2] (see [Naghdi, 1957], p. 474)

$$\begin{aligned}
e_{\alpha\beta} &= \frac{1}{2}\left(u_{\alpha|\beta} + u_{\beta|\alpha}\right) - B_{\alpha\beta}u_3, \\
\gamma_\alpha &= \delta_\alpha - \beta_\alpha, \quad \gamma_3 = \delta_3, \\
\varkappa_{\beta\alpha} &= \delta_{\beta|\alpha} - B_{\alpha\beta}\delta_3 - B_\alpha^\beta\left(u_{\nu|\beta} - B_{\nu\beta}u_3\right),
\end{aligned}$$

[2] The vertical bars on these formulas stand for covariant differentiation with respect to $A_{\alpha\beta}$.

$$
\begin{aligned}
\varkappa_{3\alpha} &= \delta_{3,\alpha} + B^{\nu}_{\alpha}(\delta_{\nu} - \beta_{\nu}),\\
\gamma_{N\alpha} &= \delta_{N\alpha}, \quad \gamma_{N3} = \delta_{N3},\\
\varkappa_{N\beta\alpha} &= \delta_{N\beta|\alpha} - B_{\beta\alpha}\delta_{N3},\\
\varkappa_{N3\alpha} &= \delta_{N3,\alpha} + B^{\beta}_{\alpha}\delta_{N\beta}
\end{aligned}
\tag{5.143}
$$

for $N \geq 2$. We can relate these equations to the infinitesimal (three-dimensional) strain tensor

$$
\gamma^{*}_{ij} = \frac{1}{2}\left(\mathbf{G}_i \cdot \mathbf{u}^{*}_{,j} + \mathbf{G}_j \cdot \mathbf{u}^{*}_{,i}\right) \tag{5.144}
$$

that comes from the linearization of the covariant components of a strain measure (5.134). This can be accomplished either by linearization of (5.135) or directly from

$$
\mathbf{u}^{*} = \mathbf{u}^{*}(\theta^{\alpha}, \xi, t) = \mathbf{u}(\theta^{\alpha}, t) + \sum_{N=1}^{\infty} \xi^{N} \boldsymbol{\delta}_N(\theta^{\alpha}, t), \qquad \boldsymbol{\delta}_N = \delta_{Ni}\mathbf{A}^{i}, \tag{5.145}
$$

together with (5.144). As a result, we obtain the following equations

$$
\begin{aligned}
2\gamma^{*}_{\alpha\beta} &= 2e_{\alpha\beta} + \xi(\varkappa_{\alpha\beta} + \varkappa_{\beta\alpha}) - 2\xi^2 B^{\nu}_{\beta}B^{\mu}_{\alpha}e_{\mu\nu}\\
&\quad + \sum_{N=2}^{\infty} \xi^{N}(\varkappa_{N\alpha\beta} + \varkappa_{N\beta\alpha}) - \sum_{N=1}^{\infty} \xi^{N+1}\left[B^{\gamma}_{\beta}\varkappa_{N\gamma\alpha} + B^{\gamma}_{\alpha}\varkappa_{N\gamma\beta}\right]\\
2\gamma^{*}_{\alpha 3} &= 2\gamma^{*}_{3\alpha} = \gamma_{\alpha} + \xi(\varkappa_{3\alpha} - B^{\nu}_{\alpha}\gamma_{\nu})\\
&\quad + \sum_{N=2}^{\infty} \xi^{N-1}\left[N\gamma_{N\alpha} + \xi\varkappa_{N3\alpha} - N\xi B^{\nu}_{\alpha}\gamma_{N\nu}\right],\\
\gamma^{*}_{33} &= \gamma_3 + \sum_{N=2}^{\infty} N\xi^{N-1}\gamma_{N3}.
\end{aligned}
\tag{5.146}
$$

Notice that these expressions for γ^{*}_{ij} are not simple at all and they clearly claim for some approximative scheme that would make it possible to derive a complete theory which would be both useful and manageable. Earlier approximation schemes for the kinematics of shells were doomed to difficulties and they were not compatible with the rest of the theory. But, prior to the consideration of a complete theory that would account for all field equations and constitutive relations, we can present at this point the linearized kinematic measures which are obtained by assuming that

$$
\gamma_{Ni} = 0 \quad \text{and} \quad \varkappa_{Ni\alpha} = 0, \quad \text{for} N \geq 2. \tag{5.147}
$$

hold for (5.146). This gives that (5.145) is simplified to $\mathbf{u}^* = \mathbf{u} + \xi\boldsymbol{\delta}$, valid for sufficiently thin shells, and the kinematic measures (5.143) become

$$\begin{aligned}
e_{\alpha\beta} &= \frac{1}{2}\left(u_{\alpha|\beta} + u_{\beta|\alpha}\right) - B_{\alpha\beta}\, u_3, \quad \gamma_\alpha = \delta_\alpha - \beta_\alpha, \quad \gamma_3 = \delta_3, \\
\beta_\alpha &= -\left(u_{3,\alpha} + B^\lambda_\alpha u_\lambda\right), \\
\varkappa_{\beta\alpha} &= \varrho_{\beta\alpha} - B_{\alpha\beta}\gamma_3, \quad \varkappa_{3\alpha} = \varrho_{3\alpha} + B^\lambda_\alpha \gamma_\lambda, \\
\varrho_{\beta\alpha} &= \gamma_{\beta|\alpha} - \bar{\varkappa}_{\beta\alpha}, \quad \varrho_{3\alpha} = \gamma_{3,\alpha}, \\
\bar{\varkappa}_{\beta\alpha} &= \bar{\varkappa}_{\alpha\beta} = \left(u_{3|\beta\alpha} + B^\nu_{\beta|\alpha} u_\nu + B^\nu_\alpha u_{\nu|\beta} + B^\nu_\beta u_{\nu|\alpha} - B^\nu_\alpha\, B_{\nu\beta} u_3\right) \\
&= -\frac{1}{2}\left(\beta_{\alpha|\beta} + \beta_{\beta|\alpha}\right) + \frac{1}{2}\left\{B^\nu_\alpha\left(e_{\nu\beta} + \gamma_{[\nu\beta]}\right) + B^\nu_\beta\left(e_{\nu\alpha} + \gamma_{[\nu\alpha]}\right)\right\}, \\
\gamma_{[\alpha\ \beta]} &= \frac{1}{2}\left(u_{\alpha|\beta} - u_{\beta|\alpha}\right).
\end{aligned} \tag{5.148}$$

As a result of this, the components of γ^*_{ij} in (5.146) now reduce to

$$\gamma^*_{\alpha\beta} = e_{\alpha\beta} + \xi\varkappa_{(\alpha\beta)} - \xi^2\chi_{\alpha\beta}, \quad \gamma^*_{\alpha 3} = \gamma^*_{3\alpha} = \gamma_\alpha + \xi\varrho_{3\alpha}, \quad \gamma^*_{33} = \gamma_3. \tag{5.149}$$

The symbol $\varkappa_{(\alpha\beta)}$ stands for the symmetric part of $\varkappa_{\alpha\beta}$, where

$$\begin{aligned}
\chi_{\alpha\beta} &= \bar{\chi}_{\alpha\beta} + B^\gamma_\alpha B^\nu_\beta e_{\lambda\nu} = \chi_{\beta\alpha}, \\
\bar{\chi}_{\alpha\beta} &= \frac{1}{2}\left(B^\gamma_\alpha \varkappa_{\gamma\beta} + B^\gamma_\beta \varkappa_{\gamma\alpha}\right).
\end{aligned} \tag{5.150}$$

In terms of physical components, the kinematic measures (5.148) are expressed as

$$\begin{aligned}
[e_{\alpha\beta}]_{F^*\otimes E^*} &= \frac{1}{2}\left([u_{\alpha|\beta}]_{F^*\otimes E^*} + [u_{\beta|\alpha}]_{E^*\otimes F^*}\right) - [B_{\alpha\beta}]_{F^*\otimes E^*}\, u_3, \\
[\gamma_\alpha]_{F^*} &= [\delta_\alpha]_{F^*} - [\beta_\alpha]_{F^*}, \quad \gamma_3 = \delta_3, \\
[\beta_\alpha]_{F^*} &= -\left([u_{3,\alpha}]_{F^*} + [B^\lambda_\alpha]_{F\otimes F^*}\, [u_\lambda]_{F^*}\right), \\
[\varkappa_{\beta\alpha}]_{E^*\otimes F^*} &= [\varrho_{\beta\alpha}]_{E^*\otimes F^*} - [B_{\alpha\beta}]_{F^*\otimes E^*}\, \gamma_3, \\
[\varkappa_{3\alpha}]_{F^*} &= [\varrho_{3\alpha}]_{F^*} + [B^\lambda_\alpha]_{E\otimes F^*}\, [\gamma_\lambda]_{E^*}, \\
[\varrho_{\beta\alpha}]_{E^*\otimes F^*} &= [\gamma_{\beta|\alpha}]_{E^*\otimes F^*} - [\bar{\varkappa}_{\beta\alpha}]_{E^*\otimes F^*}, \quad [\varrho_{3\alpha}]_{F^*} = [\gamma_{3,\alpha}]_{F^*}, \\
[\bar{\varkappa}_{\beta\alpha}]_{E^*\otimes F^*} &= -\frac{1}{2}\left([\beta_{\alpha|\beta}]_{F^*\otimes E^*} + [\beta_{\beta|\alpha}]_{F^*\otimes E^*}\right) \\
&\quad + \frac{1}{2}\Big\{[B^\nu_\alpha]_{E\otimes F^*}\left([e_{\nu\beta}]_{E^*\otimes E^*} + [\gamma_{[\nu\beta]}]_{E^*\otimes E^*}\right) \\
&\quad + [B^\nu_\beta]_{E\otimes E^*}\left([e_{\nu\alpha}]_{E^*\otimes F^*} + [\gamma_{[\nu\alpha]}]_{E^*\otimes F^*}\right)\Big\} \\
[\gamma_{[\alpha\ \beta]}]_{F^*\otimes E^*} &= \frac{1}{2}\left([u_{\alpha|\beta}]_{F^*\otimes E^*} - [u_{\beta|\alpha}]_{E^*\otimes F^*}\right).
\end{aligned} \tag{5.151}$$

5.6 Equations of motion for shells

Our main concern here is to write the general equations of motion for shells in terms of physical components. Usually this is done by starting off from the well-known energy equations for three-dimensional continuous media. Then, bearing in mind that the resultants vectors and some parameters do not vary under superposed rigid body motion, we derive the basic equations for elastic shells. However, since the three-dimensional equations are very complicated in the general theory, the strategy becomes that of reducing the three-dimensional equations to a two-dimensional form and to content ourselves with equations for stress resultants and stress couples instead of the actual stresses. It is our purpose here not to be concerned with the details of the derivation procedures but to present only the resulting equations of motion in terms of tensor components.[3] We end the section by writing these equations in terms of physical components.

Let us introduce some definitions and notations for variables that appear in the derivations of field equations for shells from the three-dimensional theory of continuum mechanics. Although we shall not concern ourselves with the derivation of the equations (task that we refer the interested reader to either [Naghdi, 1957] or [Dikmen, 1982]), we notice that the basic principles to be considered for the derivation include conservations laws and invariance requirements under superposed rigid body motions.

Consider the configurations for a shell-like body $\mathcal{B}$ together with the material surface $\mathfrak{S}$ in $\mathcal{B}$ defined by the parametrization $\xi = \xi(\theta^\alpha)$ so that the surface $\xi(\theta^\alpha) = 0$ at time t in the deformed configuration is referred as s. We designate as S, S^-, and S^+ the initial surfaces in the reference configuration of $\mathcal{B}$ which become the surfaces s, s^-, and s^+ at time t (thus, in the deformed configuration) and the initial values of $\mathbf{a}_i$, $a_{\alpha\beta}$, $b_{\alpha\beta}$, and a by $\mathbf{A}_i$, $A_{\alpha\beta}$, $B_{\alpha\beta}$, and A, respectively. Let the three-dimensional vector fields per unit mass $\mathbf{f} = \mathbf{f}(\theta^\alpha, t)$ and $\boldsymbol{l} = \boldsymbol{l}(\theta^\alpha, t)$ be defined on s in such a way that:

(1) if $\mathbf{f} \cdot \mathbf{v}$ is a rate of work per unit mass of s, for all arbitrary velocity fields $\mathbf{v}$, then $\mathbf{f}$ is called *an assigned force* vector per unit mass of s;

(2) if $\boldsymbol{l} \cdot \boldsymbol{\omega}$ is a rate of work per unit mass of s, for all arbitrary velocity fields $\boldsymbol{\omega}$, then $\boldsymbol{l}$ is called *an assigned director couple* per unit mass s.

We define the vectors fields $\overline{\mathbf{f}}$ and $\overline{\boldsymbol{l}}$ as the following differences in terms of the acceleration vector $\dot{\mathbf{v}}$ of the surface and the inertia term $\dot{\boldsymbol{\omega}}$ due to the

[3] For details of the derivations, see, for instance, Naghdi [Naghdi, 1972].

director velocity:

$$\bar{\mathbf{f}} = \mathbf{f} - \dot{\mathbf{v}} \qquad \text{and} \qquad \bar{\boldsymbol{l}} = \boldsymbol{l} - \dot{\boldsymbol{\omega}}.$$

Following the developments from the energy equation, we get that the following variables and parameters that eventually appear in power series expansions are invariant under superposed rigid motions:

$\mathbf{M}^N = \mathbf{M}^N(\theta^\alpha, t; \mathbf{v})$: Stress-resultant couple vectors of order $N \geq 0$.
$\mathbf{M}^{n\alpha}$: Stress couples of order $n \geq 0$.
$M^{\alpha i}$: Components of the contact director couple $\mathbf{M}$.
$\mathbf{N} = \mathbf{N}(\theta^\alpha, t; \mathbf{v})$: Stress-resultant force.
$\bar{\boldsymbol{l}}^n$: Body force resultants of order $n \geq 0$.
$\mathbf{m}$: Surface director couple measured per unit area in the deformed configuration .
$\mathbf{m}^n$: Shear stress resultants of order $n \geq 0$.

Then, the energy equations lead to the formulas

$$\mathbf{N}^{\alpha}_{|\alpha} + \rho\bar{\mathbf{f}} = \mathbf{0}, \tag{5.152}$$

$$\mathbf{M}^{\alpha\beta}{}_{|\alpha} + \rho\bar{\boldsymbol{l}}^n - \mathbf{m}^n = \mathbf{0}, \quad n \geq 1, \tag{5.153}$$

where ρ is the mass density per unit area of the surface and $\mathbf{M}^{n\alpha}_{|\alpha}$ means the covariant differentiation of $\mathbf{M}^{n\alpha}$ with respect to the first fundamental form of the surface. From (5.152) and (5.153) we can derive the equations of motion in component form as being

$$N^{\alpha\beta}_{\ \ |\alpha} - b^{\alpha}_{\beta} N^{\alpha 3} + \rho\bar{f}^{\beta} = 0, \tag{5.154}$$

$$N^{\alpha 3}_{|\alpha} + b_{\alpha\beta} N^{\alpha\beta} + \rho\bar{f}^{3} = 0, \tag{5.155}$$

$$M^{N\alpha\beta}_{\ \ \ |\alpha} - b^{\beta}_{\alpha} M^{N\alpha 3} + \rho\bar{l}^{N\beta} = m^{N\beta}, \tag{5.156}$$

$$M^{N\alpha 3}_{\ \ \ |\alpha} + b_{\alpha\beta} M^{N\alpha\beta} + \rho\bar{l}^{N3} = m^{N3}, \quad (N \geq 1). \tag{5.157}$$

On the other hand, taking:

(1) $N^{\alpha\beta}$ and $N^{\alpha 3}$ to stand for surface tensors under the transformation of surface coordinates (they will be explicitly defined in Subsection 5.6.1),
(2) λ to stand for the unit tangent vector to a curve on the surface in the deformed configuration, and
(3) $b_{\alpha\beta}$ to stand for the second fundamental form of the surface in the deformed configuration,

one can deduce the following set of equations:

$$N'^{\beta\alpha} = N'^{\alpha\beta} = N^{\alpha\beta} - \sum_{N=1}^{\infty} \left(m^{N\alpha} d_{N\cdot}^{\ \beta} + M^{N\gamma\alpha} \lambda_{N\cdot\gamma}^{\ \beta} \right), \tag{5.158}$$

$$N^{\alpha 3} + \sum_{N=1}^{\infty} \left(m^{N3} d_{N\cdot}^{\ \alpha} - m^{N\alpha} d_{N\cdot}^{\ 3} \right) + \sum_{N=1}^{\infty} \left(M^{N\gamma 3} \lambda_{N\cdot\gamma}^{\ \alpha} - M^{N\gamma\alpha} \lambda_{N\cdot\gamma}^{\ 3} \right) = 0, \tag{5.159}$$

and, for $n \geq 1$,

$$M'^{n\alpha\beta} = M'^{n\beta\alpha} = M^{n\alpha\beta} - \sum_{N=1}^{\infty} \left(\frac{N}{N+n} m^{N+n\alpha} d_{N\cdot}^{\ \beta} + M^{N+n\gamma\alpha} \lambda_{N\cdot\gamma}^{\ \beta} \right)$$
$$+ M^{N\alpha 3} + \sum_{N=1}^{\infty} \frac{N}{N+n} \left(m^{N+n3}\, d_{N\cdot}^{\ \alpha} - m^{N+n\alpha}\, d_{N\cdot}^{\ 3} \right)$$
$$+ \sum_{N=1}^{\infty} \left(M^{N+n\gamma 3} \lambda_{\cdot\gamma}^{\alpha} - M^{N+n\gamma\alpha}\, \lambda_{N\ \cdot\gamma}^{\ 3} \right) = 0. \tag{5.160}$$

It is very important to observe that, as we can see, the equations of motion (5.154) through to (5.160) consist in an infinite system of exact equations in an infinite number of unknowns, and this high degree of complexity requires a suitable approximation in order that the system of equations can be effectively tackled. The approximation is carried out by assuming

$$M^{N\alpha i} = m^{Ni} = 0, \quad \text{for } N \geq 2, \tag{5.161}$$

and specifying that

$$\bar{l}^{Ni} = 0, \quad \text{for } N \geq 2, \tag{5.162}$$

which is usually satisfied only approximately. Hence, the approximation means that we content ourselves with only the information contained in the moments of order zero and one. With this approximation, the infinite set of equations is replaced by the following reduced finite set, which are formally the equations of motion for a Cosserat surface when the surfaces S and s are identified:

$$N^{\alpha\beta}_{\ \ |\alpha} - b^{\beta}_{\alpha}N^{\alpha 3} + \rho^{\beta}\bar{f}^{\beta} = 0, \tag{5.163}$$

$$N^{\alpha 3}_{\ \ |\alpha} + b_{\alpha\beta}N^{\alpha\beta} + \rho\bar{f}^{3} = 0, \tag{5.164}$$

$$M^{\alpha\beta}_{\ \ |\alpha} - b^{\beta}_{\alpha}M^{\alpha 3} + \rho\bar{l}^{\beta} = m^{\beta}, \tag{5.165}$$

$$M^{\alpha\beta}_{\ \ |\alpha} + b_{\alpha\beta}M^{\alpha\beta} + \rho\bar{l}^{3} = m^{3}, \tag{5.166}$$

$$N'^{\alpha\beta} = N'^{\beta\alpha} = N^{\alpha\beta} - m^{\alpha}d^{\beta} - M^{\gamma\alpha}\lambda^{\ \beta}_{.\gamma}, \tag{5.167}$$

$$N^{\alpha 3} + m^{3}d^{\alpha} - m^{\alpha}d^{3} + M^{\gamma 3}\lambda^{\ \alpha}_{.\gamma} - M^{\gamma\alpha}\lambda^{\ 3}_{.\gamma} = 0. \tag{5.168}$$

Consistent with the above approximations, we also take

$$\bar{\mathbf{f}} = \mathbf{f} - \dot{\mathbf{v}} \quad \text{and} \quad \bar{\boldsymbol{l}} = \boldsymbol{l} - k^{11}\dot{\boldsymbol{\omega}}, \tag{5.169}$$

where

$$\rho k^{11}a^{1/2} = \int_{\alpha}^{\beta} \rho^{*}g^{1/2}\xi^{2}d\xi.$$

As a consequence of these approximations, equations (5.160) reduce to

$$M'^{1\alpha\beta} = M^{\alpha\beta} = M^{\beta\alpha}, \tag{5.170}$$

$$M^{1\alpha 3} = M^{\alpha 3} = 0. \tag{5.171}$$

In an exact theory, these equations are proper identities. However, they can only be satisfied in some special cases in an approximate theory. As for the equations (5.163)–(5.168), it is worthwhile to mention that they are formally the same as the equations of motion for a Cosserat surface from the literature if the surfaces S and s are identified.

Therefore, the **general approximate nonlinear equations of motion in terms of tensor components** are specified by (5.163)–(5.168) together with (5.170) and we can write them down again organized as the system (δ) below:

$$(\delta)\quad \begin{cases} N^{\alpha\beta}_{\ \ |\alpha} - b^{\beta}_{\alpha}N^{\alpha 3} + \rho^{\beta}\bar{f}^{\beta} = 0, \quad {N^{\alpha 3}}_{|\alpha} + b_{\alpha\beta}N^{\alpha\beta} + \rho\bar{f}^{3} = 0, \\ M^{\alpha\beta}_{\ \ |\alpha} - b^{\beta}_{\alpha}M^{\alpha 3} + \rho\bar{l}^{\beta} = m^{\beta}, \quad {M^{\alpha 3}}_{|\alpha} + b_{\alpha\beta}M^{\alpha\beta} + \rho\bar{l}^{3} = m^{3}, \\ N'^{\alpha\beta} = N'^{\beta\alpha} = N^{\alpha\beta} - m^{\alpha}d^{\beta} - M^{\gamma\alpha}\lambda^{\ \beta}_{.\gamma}, \\ N^{\alpha 3} + m^{3}d^{\alpha} - m^{\alpha}d^{3} + M^{\gamma 3}\lambda^{\ \alpha}_{.\gamma} - M^{\gamma\alpha}\lambda^{\ 3}_{.\gamma} = 0, \\ M'^{1\alpha\beta} = M^{\alpha\beta} = M^{\beta\alpha}, \quad M^{1\alpha 3} = M^{\alpha 3} = 0. \end{cases}$$

In terms of physical components, these general approximate equations of motion become the system (ψ):

$$
(\psi)\quad \begin{cases}
\left[N^{\alpha\beta}_{\ \ |\alpha}\right]_{f\otimes e\otimes f^*} - [b^{\beta}_{\alpha}]_{f^*\otimes e}[N^{\alpha 3}]_f + \rho^{\beta}[\bar{f}^{\beta}]_e = 0,\\
\left[N^{\alpha 3}_{\ \ |\alpha}\right]_{f\otimes f^*} + [b_{\alpha\beta}]_{f^*\otimes e^*}[N^{\alpha\beta}]_{f\otimes e} + \rho \bar{f}^{\,3} = 0,\\
\left[M^{\alpha\beta}_{\ \ |\alpha}\right]_e - [b^{\beta}_{\alpha}]_{f^*\otimes e}[M^{\alpha 3}]_f + \rho[\bar{l}^{\beta}]_e = [m^{\beta}]_e,\\
M^{\alpha 3}_{\ \ |\alpha} + [b_{\alpha\beta}]_{f^*\otimes e^*}[M^{\alpha\beta}]_{f\otimes e} + \rho\bar{l}^3 = m^3,\\
[N'^{\alpha\beta}]_{f\otimes e} = [N'^{\beta\alpha}]_{e\otimes f} = [N^{\alpha\beta}]_{f\otimes e} - [m^{\alpha}]_f[d^{\beta}]_e\\
\qquad -[M^{\gamma\alpha}]_{e\otimes f}[\lambda^{\beta}_{.\gamma}]_{e\otimes e^*},\\
[N^{\alpha 3}]_f + m^3[d^{\alpha}]_f - [m^{\alpha}]_f d^3 + [M^{\gamma 3}]_e[\lambda^{\alpha}_{.\gamma}]_{f\otimes e*}\\
\qquad -[M^{\gamma\alpha}]_{e\otimes f}[\lambda^{3}_{.\gamma}]_{e^*} = 0.
\end{cases}
$$

The equations in terms of physical components in system (ψ) above are homogeneous in the sense that the terms in each one of the equations have the same physical dimensions.

One way to determine the solution of the problem consists in solving the system (ψ). Another way is to solve the system (δ) and substitute the result in (ϕ):

$$
(\phi)\quad \begin{cases}
[N^{\alpha\beta}]_{f\otimes e} = \dfrac{\sqrt{a_{\beta\beta}}}{\sqrt{a^{\alpha\alpha}}}N^{\alpha\beta}, \quad [N^{\alpha 3}]_f = \dfrac{1}{\sqrt{a^{\alpha\alpha}}}N^{\alpha 3},\\
[M^{\alpha\beta}]_{f\otimes e} = \dfrac{\sqrt{a_{\beta\beta}}}{\sqrt{a^{\alpha\alpha}}}M^{\alpha\beta}, \quad [M^{\alpha 3}]_f = \dfrac{1}{\sqrt{a^{\alpha\alpha}}}M^{\alpha 3},\\
[M^{\gamma 3}]_f = \sqrt{a_{\gamma\gamma}}M^{\gamma 3}, \quad [M^{\gamma\alpha}]_{e\otimes f} = \dfrac{\sqrt{a_{\gamma\gamma}}}{\sqrt{a^{\alpha\alpha}}}M^{\gamma\alpha},\\
\left[N^{\alpha\beta}_{\ \ |\alpha}\right]_{f\otimes e\otimes f^*} = \sqrt{a_{\beta\beta}}N^{\alpha\beta}_{\ \ |\alpha}, \quad \left[M^{\alpha\beta}_{\ \ |\alpha}\right]_{f\otimes e\otimes f^*} = \sqrt{a_{\beta\beta}}M^{\alpha\beta}_{\ \ |\alpha},\\
[b_{\alpha\beta}]_{f^*\otimes e^*} = \dfrac{\sqrt{a^{\alpha\alpha}}}{\sqrt{a_{\beta\beta}}}b_{\alpha\beta}, \quad [b_{\alpha}^{\ \cdot\beta}]_{f^*\otimes e} = \sqrt{a^{\alpha\alpha}}\sqrt{a_{\beta\beta}}b_{\alpha}^{\ \cdot\beta},\\
[b^{\alpha}_{\ \gamma}]_{f\otimes e^*} = \dfrac{1}{\sqrt{a^{\alpha\alpha}}\sqrt{a_{\gamma\gamma}}}b^{\alpha}_{\ \gamma}, \quad [b_{\gamma}^{\ \cdot\beta}]_{e^*\otimes e} = \dfrac{\sqrt{a_{\beta\beta}}}{\sqrt{a_{\gamma\gamma}}}b_{\gamma}^{\ \cdot\beta},\\
[\bar{f}^{\beta}]_e = \sqrt{a_{\beta\beta}}\bar{f}^{\beta}, \quad [\bar{l}^{\beta}]_e = \sqrt{a_{\beta\beta}}\bar{l}^{\beta},\\
[k^{\beta}]_e = \sqrt{a_{\beta\beta}}k^{\beta}, \quad [k^{\alpha}]_f = \dfrac{1}{\sqrt{a^{\alpha\alpha}}}k^{\alpha}.
\end{cases}
$$

Finally, we note that equations (5.152) and (5.153) were used in Dikmen [Dikmen, 1982] to derive the equations of motion for shells, where it was observed that the results coincided with those obtained in Naghdi [Naghdi, 1972] by requiring the invariance of the balance of energy under rigid body translation and rigid body rotation. Therefore, considering the motion of the shell and some invariance requirements under superposed rigid body motion, one can deduce another set of equations of motion to complement (5.152) and (5.153), namely,

$$\mathbf{M}^{na} \times \mathbf{a}_\alpha + \sum_{N=1}^{\infty} \left(\frac{N}{N+n} \mathbf{m}^{N+n} \times \mathbf{d}_N + \mathbf{M}^{N+n\alpha} \times \mathbf{d}_{N,\alpha} \right) = \mathbf{0}, \qquad (5.172)$$

for $n \geq 0$. Recall that the $\mathbf{a}_i$'s are the basis vectors and the $\mathbf{d}_N$'s are the vector fields occurring in a representation of the position vector $\mathbf{p}$ of the place in a body occupied by the material point in the deformed configuration. For $n = 0$, (5.172) is written as

$$\mathbf{N}^{na} \times \mathbf{a}_\alpha + \sum_{N=1}^{\infty} \left(\mathbf{m}^{N} \times \mathbf{d}_N + \mathbf{M}^{N+\alpha} \times \mathbf{d}_{N,\alpha} \right) = \mathbf{0}. \qquad (5.173)$$

Thus, equations (5.153) together with (5.172) are the vector equations for the shell motion, with equation (5.172) being a consequence of the symmetry of the stress tensor. We can refer all the vector fields and resultants to the basis vectors $\mathbf{a}_i$ to yield

$$\mathbf{N}^{\alpha} = N^{\alpha i}\mathbf{a}_i, \qquad \mathbf{M}^{N\alpha} = M^{N\alpha i}\mathbf{a}_i \qquad (5.174)$$

and

$$\mathbf{m}^{N} = m^{Ni}\mathbf{a}_i. \qquad (5.175)$$

Similarly,

$$\bar{\mathbf{f}} = \bar{\mathbf{f}}^{o} = \bar{f}^{i}\mathbf{a}_i, \quad \bar{\mathbf{l}} = \bar{\mathbf{l}}^{1} = \bar{l}^{i}\mathbf{a}_i, \quad \bar{\mathbf{l}}^{N} = \bar{l}^{Ni}\mathbf{a}_i, \quad (N \geq 2). \qquad (5.176)$$

5.6.1 *Linearized equations of motion*

In order to obtain a linearized version of the approximate equations of motion (5.163)–(5.170), we suppose that the continuum is initially free of stress and that all kinematic measures and their derivatives with respect to the surface coordinates as well as with respect to time are terms or group of terms of order[4] $\mathcal{O}(\varepsilon)$. The equations of equilibrium (and motion) for shells in general coordinates were derived via the direct method presented in [Synge and Chien, 1941]. Another derivation by the same method, carried out in [Ericksen and Truesdell, 1958], turned out to be shorter and more elegant. As early as 1888, [Love, 1888] derived the equations of equilibrium for shells using a method that was only in part direct. As an alternative to the direct method, the derivation of the equations of motion from the three-dimensional equations involves the concepts of stress resultants and stress couples and is accomplished when the coordinate curves on the middle surface are taken as lines of curvature. Love approached the problem in two steps. First, the stress resultants $\{N^{\alpha\beta}, V^{\alpha}\}$ and the stress couples $M^{\alpha\beta}$ (measured per unit length of coordinate curves on the surface S) were defined by the following integrals,

$$N^{\alpha\beta} = \int_{-k/2}^{k/2} \mu\tau^{\alpha\beta}\mu_{\gamma}^{\beta} d\zeta, \qquad V^{\alpha} = \int_{-k/2}^{k/2} \mu\tau^{\alpha 3} d\zeta, \tag{5.177}$$

$$M^{\alpha\beta} = \int_{-k/2}^{k/2} \mu\tau^{\alpha\gamma}\mu_{\gamma}^{\beta}\zeta d\zeta, \qquad M^{\alpha 3} = \int_{-k/2}^{k/2} \mu\tau^{\alpha 3}\zeta d\zeta, \tag{5.178}$$

$$V^{3} = \int_{-k/2}^{k/2} \mu\left(\tau^{33} - B_{\alpha\beta}\,\tau^{\alpha\gamma}\mu_{\gamma}^{\beta}\zeta\right) d\zeta, \tag{5.179}$$

together with similar definitions for the load resultants. Then, instead of deriving the equilibrium equations for shells from the three-dimensional equations, he proceeded as follows. He considered the equilibrium of an element of the curved shell (actually, its middle surface) under the action of the stress resultants and stress couples. Each one of these actions were taken per unit length of curves on the middle surface and of the load resultant per unit area of the middle surface. Thus, it was only the second step of the procedure that was done by the direct method. In [Reissner, 1941] one can find a vectorial treatment of Love's derivation in terms of lines of curvature coordinates. The definition of stress resultants and stress couples in terms of the three-dimensional stress tensor (corresponding to the resultants in (5.178) and (5.179)) but in lines of curvature coordinates for physical components of $N^{\alpha\beta}$, $M^{\alpha\beta}$ and V^{α}) is due to Love. The definitions of the resultants in general coordinates corresponding to $N^{\alpha\beta}$, $M^{\alpha\beta}$, and V^{α} in (5.180) below were given in [Zerna, 1968].

[4] We recall that the notation $A = B + \mathcal{O}(\varepsilon^k)$ means that the order symbol $\mathcal{O}(\varepsilon^k)$ is in the place of the terms or group of terms for which there exists a real number $K > 0$, which is independent of ε and of the functions in question together with their derivatives, such that $|\mathcal{O}(\varepsilon^k)| < K\varepsilon^k$ as $\varepsilon \to 0$.

The earliest derivation which can be regarded as fully obtained from the three-dimensional equations appeared in [Novozhilov, 1970] and [Novozhilov and Finkelshtein, 1943]. This derivation, carried out in lines of curvature coordinates and in terms of physical components of the resultants, is accomplished by integration of the three-dimensional differential equations of equilibrium across the thickness of the shell. In these equations, the derivatives are independent of any kinematic assumptions and no approximation is effected. An exposition of the derivation carried out in [Novozhilov and Finkelshtein, 1943] is also given in [Truesdell and Toupin, 1960].

A linearized procedure shows that a linearized version of the equations of motion is obtained by referring all tensors in these equations to the initial underformed surface (reference configuration), covariant differentiation being with respect to $A_{\alpha\beta}$ and replacing $b_{\alpha\beta}$, d_{Ni}, $\lambda_{Ni\alpha}$ and ρ by $B_{\alpha\beta}, D_{Ni}, \Lambda_{Ni\alpha}$, and ρ_o, respectively. A derivation from the three-dimensional theory in general coordinates resulting in the equilibrium equations above involving $N^{\alpha\beta}$, $M^{\alpha\beta}$, and V^{α} can be found in [Naghdi, 1972].

With a partly different notation, we now confine our attention to three linearized versions of the approximate equations of motion (5.163)–(5.168). Let $\overline{F}^i$ and $\overline{L}^i$ denote the components of $\overline{f}$ and $\overline{\ell}$ referred to the basis vectors $\overline{A}_i$, that is, $\overline{F}^i = \overline{f}A_i$ and $\overline{L}^i = \overline{\ell}A_i$.

1) For $\mathbf{D} = \mathbf{A}_3$, $\Lambda_{\beta\alpha} = -B_{\alpha\beta}$, and $\Lambda_{3\alpha} = 0$ and the stress resultants and stress couples defined by (5.177), (5.178), and (5.179), the linearized equations of motion become

$$\begin{aligned}
N^{\alpha\beta}{}_{|\alpha} - B^{\beta}_{\alpha}V^{\alpha} + \rho_o\overline{F}^{\beta} &= 0,\\
V^{\alpha}_{|\alpha} + B_{\alpha\beta}N^{\alpha\beta} + \rho_o\overline{F}^{3} &= 0,\\
M^{\alpha\beta}{}_{|\alpha} + \rho_o\overline{L}^{\beta} &= V^{\beta},\\
M^{\alpha 3}{}_{|\alpha} + \rho_o\overline{L}^{3} &= V^{3},\\
N'^{\alpha\beta} = N'^{\beta\alpha} &= N^{\alpha\beta} + M^{\gamma\alpha}B^{\beta}_{\gamma},\\
V^{\alpha} = N^{\alpha 3} = m^{\alpha} + B^{\alpha}_{\gamma}M^{\gamma 3}, &\quad V^{3} = m^{3} - B_{\alpha\beta}M^{\alpha\beta}.
\end{aligned} \tag{5.180}$$

This set of linearized equations of motion rewritten in terms of physical components is[5]

[5]We note that, for scalar equations, the components coincide with the physical components (projected in the direction of the unit normal).

$$\left[N^{\alpha\beta}_{\ \ |\alpha}\right]_{F\otimes E\otimes F^*} - [B^{\beta}_{\alpha}]_{E\otimes F^*}[V^{\alpha}]_F + \rho^{\beta}[\overline{F}^{\beta}]_E = 0,$$

$$\left[V^{\alpha}_{|\alpha}\right]_{F\otimes F^*} + [B_{\alpha\beta}]_{F^*\otimes E^*}\left[N^{\alpha\beta}\right]_{F\otimes E} + \varrho_o\overline{F}^3 = 0,$$

$$\left[M^{\alpha\beta}_{\ \ |\alpha}\right]_{F\otimes E\otimes F^*} + \varrho_o\left[\overline{L}^{\beta}\right]_E = \left[V^{\beta}\right]_E,$$

$$M^{\alpha 3}_{\ \ |\alpha} + \rho_o\overline{L}^3 = V^3 \quad \text{(scalar equation)},$$

$$\left[N'^{\alpha\beta}\right]_{F\otimes E} = \left[N^{\alpha\beta}\right]_{F\otimes E} + \left[B^{\beta}_{\gamma}\right]_{F^*\otimes E}\left[M^{\gamma\alpha}\right]_{F\otimes E},$$

$$[V^{\alpha}]_F = \left[N^{\alpha 3}\right]_F = [m^{\alpha}]_F + \left[B^{\alpha}_{\gamma}\right]_{F^*\otimes F}\left[M^{\gamma 3}\right]_E,$$

$$V^3 = m^3 - [B_{\alpha\beta}]_{F^*\otimes E^*}\left[M^{\alpha\beta}\right]_{F\otimes E}.$$

2) For $\mathbf{D} = D\mathbf{A}_3$, $\Lambda_{\nu\alpha} = -B_{\nu\alpha}$, and $\Lambda_{3\alpha} = D_{,\alpha}$ and the stress resultants and stress couples defined by

$$\hat{N}^{\alpha\beta} = \int_{-k/2}^{k/2} \mu\hat{\tau}^{\alpha\beta}\mu^{\beta}_{\gamma}d\zeta, \quad V^{\alpha} = \int_{-k/2}^{k/2} \mu\hat{\tau}^{\alpha 3}d\varsigma, \tag{5.181}$$

$$\hat{M}^{\alpha\beta} = \int_{-k/2}^{k/2} \mu\hat{\tau}^{\alpha\gamma}\mu^{\beta}_{\gamma}\varsigma d\zeta, \quad \hat{M}^{\alpha 3} = \int_{-k/2}^{k/2} \mu\hat{\tau}^{\alpha 3}\zeta d\zeta, \tag{5.182}$$

$$V^3 = \int_{-k/2}^{k/2} \mu\left(\hat{\tau}^{33} - B_{\alpha\beta}\,\hat{\tau}^{\alpha\gamma}\mu^{\beta}_{\gamma}\zeta\right)d\zeta, \tag{5.183}$$

we have

$$\hat{N}^{\alpha\beta}_{\ \ |\alpha} - B^{\beta}_{\alpha}V^{\alpha} + \rho_o\bar{F}^{\beta} = 0, \qquad V^{\alpha}_{|\alpha} + B_{\alpha\beta}\hat{N}^{\alpha\beta} + \rho_o\bar{F}^3 = 0,$$

$$\hat{M}^{\alpha\beta}_{\ \ |\alpha} + \rho_o\hat{L}^{\beta} = V^{\beta}, \qquad \hat{M}^{\alpha 3}_{\ \ |\alpha} + \rho_o\hat{L}^3 = V^3,$$

and

$$N'^{\alpha\beta} = N'^{\beta\alpha} = \hat{N}^{\alpha\beta} + \hat{M}^{\gamma\alpha}B^{\beta}_{\gamma}, \tag{5.184}$$

$$V^{\alpha} = D\,m^{\alpha} + B^{\alpha}_{\gamma}\hat{M}^{\gamma 3} + \hat{M}^{\beta\alpha}\frac{D_{,\beta}}{D}, \tag{5.185}$$

$$V^3 = D\,m^3 - B_{\alpha\beta}\hat{M}^{\alpha\beta} + \hat{M}^{\alpha 3}\frac{D_{,\alpha}}{D}, \tag{5.186}$$

$$\hat{L}^{\beta} = D\bar{L}^{\beta}, \quad \hat{L}^3 = D\bar{L}^3. \tag{5.187}$$

In terms of physical components, the equations are as follows:

$$\left[\hat{N}^{\alpha\beta}_{\ \ |\alpha}\right]_{F\otimes E\otimes F^*} - [B^{\beta}_{\alpha}]_{E\otimes F^*}[V^{\alpha}]_F + \rho^{\beta}[\overline{F}^{\beta}]_E = 0,$$

$$\left[V^{\alpha}_{|\alpha}\right]_{F\otimes F^*} + [B_{\alpha\beta}]_{F^*\otimes E^*}\,[N^{\alpha\beta}]_{F\otimes E} + \varrho_o\overline{F}^3 = 0,$$

$$\hat{M}^{\alpha 3}_{\ \ |\alpha} + \rho_o\hat{L}^3 = V^3 \quad \text{(scalar equation)},$$

$$\left[N'^{\alpha\beta}\right]_{F\otimes E} = \left[\hat{N}^{\alpha\beta}\right]_{F\otimes E} + [B^{\beta}_{\gamma}]_{F^*\otimes E}\;[M^{\gamma\alpha}]_{F\otimes E},$$

$$[V^{\alpha}]_E = D\;[m^{\alpha}]_E + [B^{\alpha}_{\gamma}]_{F^*\otimes E}\,[M^{\gamma 3}]_F, + \left[\hat{M}^{\alpha\beta}\right]_{F\otimes E}\frac{[D_{,\beta}]_{F^*}}{D},$$

$$V^3 = D\,m^3 - [B_{\alpha\beta}]_{F^*\otimes E^*}\left[\hat{M}^{\alpha\beta}\right]_{F\otimes E} + \left[\hat{M}^{\alpha 3}\right]_E\frac{[D_{,\alpha}]_{E^*}}{D},$$

$$\left[\hat{L}^{\beta}\right]_F = D\;[\bar{L}^{\beta}]_F,\quad \hat{L}^3 = D\bar{L}^3.$$

3) For $\mathbf{D} = D\mathbf{A}_3$, $\Lambda_{\beta\alpha} = -B_{\alpha\beta}D$, and $\Lambda_{3\alpha} = D_{,\alpha}$ and the stress resultants and stress couples defined by

$$\begin{aligned}
N^{\alpha\beta} &= \int_{-\beta}^{\beta}\nu\tau^{\alpha\gamma}\nu^{\beta}_{\gamma}d\xi, & N^{\alpha 3} &= \int_{-\beta}^{\beta}\nu(\xi D_{,\beta}\tau^{\alpha\beta} + D\tau^{\alpha 3})d\xi,\\
M^{\alpha\beta} &= \int_{-\beta}^{\beta}\nu\tau^{\alpha\gamma}\nu^{\beta}_{\gamma}\xi d\xi, & M^{\alpha 3} &= \int_{-\beta}^{\beta}\nu(\xi D_{,\beta}\tau^{\alpha\beta} + D\tau^{\alpha 3})\xi d\xi,\\
m^{\alpha} &= \int_{-\beta}^{\beta}\nu\tau^{3\gamma}\nu^{\alpha}_{\gamma}d\xi, & m^3 &= \int_{-\beta}^{\beta}\nu(\xi D_{,\alpha}\tau^{3\alpha} + D\tau^{33})d\xi,
\end{aligned} \tag{5.188}$$

we have the following linearized equations:

$$\begin{aligned}
N^{\alpha\beta}_{\ \ |\alpha} - B^{\beta}_{\alpha}N^{\alpha 3} + \rho_o\bar{F}^{\beta} &= 0,\\
N^{\alpha 3}_{\ \ |\alpha} + B_{\alpha\beta}N^{\alpha\beta} + \rho_o\bar{F}^3 &= 0,\\
M^{\alpha\beta}_{\ \ |\alpha} - B^{\beta}_{\alpha}M^{\alpha 3} + \rho^{\beta}_o\bar{L}^{\beta} &= m^{\beta},\\
M^{\alpha 3}_{\ \ |\alpha} + B_{\alpha\beta}M^{\alpha\beta} + \rho_o\bar{L}^3 &= m^3,
\end{aligned} \tag{5.189}$$

and

$$N'^{\alpha\beta} = N'^{\beta\alpha} = N^{\alpha\beta} + D\, M^{\gamma\alpha} B^{\beta}_{\alpha},$$
$$N^{\alpha 3} - D\, m^{\alpha} - D\, M^{\gamma 3} B^{\alpha}_{\gamma} - M^{\gamma\alpha} D_{,\gamma} = 0, \tag{5.190}$$

$$N'^{\alpha\beta} = N'^{\beta\alpha} = N^{\alpha\beta} - m^{\alpha} D^{\beta} - M^{\gamma\alpha} \Lambda^{\beta}_{.\gamma},$$
$$N^{\alpha 3} + m^3 D^{\alpha} - m^{\alpha} D^3 + M^{\gamma 3} \Lambda^{\alpha}_{.\gamma} - M^{\gamma\alpha} \Lambda^{3}_{.\gamma} = 0. \tag{5.191}$$

with $\overline{\mathbf{F}}^i = \overline{F}^i \mathbf{A}_i$ and $\overline{\mathbf{L}} = \overline{L}\mathbf{A}_i$. Writing these linearized equations in terms of physical components, we have

$$[N^{\alpha\beta}_{\ |\alpha}]_{F\otimes E\otimes F^*} - [B^{\beta}_{\alpha}]_{F^*\otimes E}[N^{\alpha 3}]_F + \varrho_o[\overline{F}^{\beta}]_E = 0,$$
$$[N^{\alpha 3}_{\ |\alpha}]_{F\otimes F^*} + [B_{\alpha\beta}]_{F^*\otimes E}[N^{\alpha\beta}]_{F\otimes E} + \varrho_o\overline{F}^3 = 0,$$
$$[M^{\alpha\beta}_{\ |\alpha}]_{F^*\otimes E\otimes F^*} - [B^{\beta}_{\alpha}]_{F^*\otimes E}[M^{\alpha 3}]_F + \varrho_o[\overline{L}^{\beta}]_E = [m^{\beta}]_E,$$
$$[M^{\alpha 3}_{\ |\alpha}]_{F\otimes F^*} + [B_{\alpha\beta}]_{F^*\otimes E^*}[M^{\alpha\beta}]_{F\otimes E} + \varrho_o\overline{L}^3 = m^3,$$
$$[N'^{\beta\alpha}]_{F\otimes E} = [N'^{\beta\alpha}]_{E\otimes F} = [N^{\alpha\beta}]_{F\otimes E} + D\,[M^{\gamma\alpha}]_{E\otimes F}[B^{\beta}_{\gamma}]_{E\otimes E^*},$$
$$[N^{\alpha 3}]_F + D\,[m^{\alpha}]_F - D\,[M^{\gamma 3}]_E[B^{\alpha}_{\gamma}]_{F\otimes E^*} - [M^{\gamma\alpha}]_{E\otimes F}[D_{,\gamma}]_{E^*} = 0.$$

5.7 Constitutive equations for thermoelastic shells

The kinematic relations together with equations of motion need to be supplemented with constitutive equations that take into account the material properties of the elastic bodies. The resulting set of equations and appropriate boundary and initial conditions suffice to establish the well-posedness of the problem.

In the three-dimensional theory of (nonpolar) elastic materials, the Helmholtz free energy density per unit of mass ψ^* is given by the following formula in terms of the internal energy ε^*, the temperature θ^*, and the entropy η^*, all being densities per unit mass (see Reddy [Reddy, 2013]):

$$\psi^* = \varepsilon^* - \theta^* \eta^*. \tag{5.192}$$

If $\psi^* = \psi^*(\gamma^*_{ij}, \theta^*)$, that is, ψ^* depends only on the strains γ^*_{ij} and the temperature θ^*, we have that

$$\eta^* = -\frac{\partial \psi^*}{\partial \theta^*}. \tag{5.193}$$

In the cases of adiabatic–isentropic deformation or isothermal deformation[6] with reversible heat transfer, the stress tensor (density per unit of mass) τ_{ij} is obtained by the differentiation

$$\tau_{ij} = \frac{\partial \psi^*}{\partial \gamma^*_{ij}}, \tag{5.194}$$

which can be rewritten in terms of the Cauchy stress tensor $T^{ij} = \rho^* \tau_{ij}$ as

$$T^{ij} = \rho^* \frac{\partial \psi^*}{\partial \gamma^*_{ij}}. \tag{5.195}$$

The conservation law for the mass is expressed locally as

$$\rho^* \sqrt{g} = \rho^*_o \sqrt{G}, \tag{5.196}$$

where

$$\rho^* = \rho^*(\theta^\alpha, \xi, t) \quad \text{and} \quad \rho^*_o = \rho^*(\theta^\alpha, \xi, 0)$$

are the mass densities at a point of any configuration and at the corresponding point in the reference configuration ($t = 0$), respectively, and G is the dual of g in a reference configuration.

The expressions (5.194) and (5.195) are obtained in the quite specific situations of either adiabatic and isentropic or isothermal deformations. In thesed situations, the thermal effects are ignored and the energy turns out as a strain-energy density condition $\psi^* = \psi^*(\gamma^*_{ij})$. Under such restrictions, the theory becomes purely mechanical and we say that the deformed material is *hyperelastic*. The constitutive equations for shells that we present below are obtained under the assumption that the material behaves hyperelastically.

In order to indicate the nature of the reduction to two-dimensional counterparts for the above results, we first observe the relations:

$$\frac{\partial \psi^*}{\partial \mathbf{g}_k} = \frac{\partial \psi^*}{\partial \gamma^*_{ij}} \cdot \frac{\partial \gamma^*_{ij}}{\partial \mathbf{g}_k} = \frac{\partial \psi^*}{\partial \gamma^*_{ij}} \left[\frac{1}{2} \left(\delta^k_i \mathbf{g}_j + \delta^k_j \mathbf{g}_i \right) \right] = \frac{\partial \psi^*}{\partial \gamma^*_{kj}} \mathbf{g}_j, \tag{5.197}$$

$$\frac{\partial \psi^*}{\partial \mathbf{a}_\alpha} = \frac{\partial \psi^*}{\partial \gamma^*_{\alpha j}} \mathbf{g}_j, \tag{5.198}$$

[6] An adiabatic deformation is one that occurs without exchanging heat with its environment. In an isentropic deformation, the entropy is invariant throughout the process. Isothermal deformations are those in which the temperature does not vary during the process.

$$\frac{\partial\psi^*}{\partial\mathbf{d}_N} = \frac{\partial\psi^*}{\partial\gamma^*_{3j}} N\xi^{N-1}\mathbf{g}_j, \qquad (5.199)$$
$$\frac{\partial\psi^*}{\partial\mathbf{d}^\alpha_N} = \frac{\partial\psi^*}{\partial\gamma^*_{\alpha j}}\mathbf{g}_j\xi^N.$$

The equality $\dfrac{\partial\psi^*}{\partial\mathbf{g}_k} = \dfrac{\partial\psi^*}{\partial\gamma^*_{ij}}\mathbf{g}_j$ in (5.197) is a consequence of $\psi^* = \psi^*(\gamma^*_{ij}, \theta^*)$ whenever the argument γ^*_{ij} is expressed in terms of the vectors $\mathbf{g}_i$ and $\mathbf{G}_i$ as[7]

$$\gamma^*_{ij} = \frac{1}{2}\left(g_{ij} - G_{ij}\right).$$

The equation (5.199) is obtained with the help of the series expansions of $\mathbf{g}_\alpha$ and $\mathbf{g}_3$, namely,

$$\mathbf{g}_\alpha = \mathbf{a}_\alpha + \sum_{N=1}^{\infty}\frac{\partial\mathbf{d}_N}{\partial\theta^\alpha}\xi^N \quad \text{and} \quad \mathbf{g}_3 = \sum_{N=1}^{\infty} N\mathbf{d}_N\xi^{N-1} \qquad (5.200)$$

and the series expansion of θ^*,

$$\theta^* = \varphi_o + \sum_{N=1}^{\infty}\varphi_N\xi^N$$

where $\varphi_o = \theta^*(\theta^\alpha, 0, t)$ and each φ_N is a scalar function depending on θ^α and t.

At this point, it is convenient to introduce the vectors $\mathbf{T}^i$ defined as

$$\mathbf{T}^i = \sqrt{g}T^{ij}\mathbf{g}_j.$$

Substituting the expression for T^{ij} in (5.195) in this expression, we have

$$\mathbf{T}^i = \rho^*\sqrt{g}\frac{\partial\psi^*}{\partial\gamma^*_{ij}}\mathbf{g}_j. \qquad (5.201)$$

In the derivation of the equations of motion, we have set

$$\mathbf{M}^{N\alpha}\sqrt{a} = \int_\alpha^\beta \mathbf{T}^\alpha\xi^N d\xi, \qquad (5.202)$$

$$\mathbf{m}^N\sqrt{a} = N\int_\alpha^\beta \mathbf{T}^3\xi^{N-1}d\xi, \qquad (5.203)$$

[7] There is a difference of notation between Naghdi [Naghdi, 1972] and Dikmen [Dikmen, 1982]. Here we follow the notation of Dikmen.

where the integration is over the interval $[\alpha, \beta]$ (α here not to be confused with the index α), and distinguish the particular case of $N = 0$ by setting

$$\mathbf{N}^{\alpha} = \mathbf{M}^{\alpha} = \frac{1}{\sqrt{a}} \int_{\alpha}^{\beta} \mathbf{T}^{\alpha} d\xi. \tag{5.204}$$

Let us now seek relations which link $\mathbf{N}^{\alpha}$, $\mathbf{M}^{\alpha}$, and $\mathbf{m}^{N}$ to the quantities $\mathbf{a}_{\alpha}, \mathbf{d}^{N}$ and $\mathbf{d}^{N\alpha}$. Integrating the constitutive equations (5.201) for $\mathbf{T}^{\alpha}$ with respect to the variable ξ, we have

$$\int_{\alpha}^{\beta} \mathbf{T}^{\alpha} d\xi = \int_{\alpha}^{\beta} \rho^{*} \sqrt{g} \frac{\partial \psi^{*}}{\partial \gamma_{\alpha j}^{*}} \mathbf{g}_{j} d\xi = \int_{\alpha}^{\beta} \kappa \frac{\partial \psi^{*}}{\partial \gamma_{\alpha j}^{*}} \mathbf{g}_{j} d\xi. \tag{5.205}$$

The time-independent quantity $\kappa = \kappa(\theta^{\alpha}, \xi)$ is introduced by using the conservation law expressed by (5.196) and setting

$$\kappa = \rho^{*} \sqrt{g}. \tag{5.206}$$

For convenience, we also introduce two new variables:

(1) the mass of $\rho(\theta^{\alpha}, t)$ per unit area on the middle surface determined by the vector position of the generic point can be defined by setting

$$\rho \sqrt{a} = \int_{\alpha}^{\beta} \kappa d\xi. \tag{5.207}$$

(2) the strain-energy density $\overline{\psi}$ per unit of area over the same middle surface is defined by

$$\overline{\psi} = \frac{1}{\rho \sqrt{a}} \int_{\alpha}^{\beta} \kappa \psi^{*} d\xi. \tag{5.208}$$

Inserting (5.198) and (5.208) in (5.205), we finally get

$$\mathbf{N}^{\alpha} = \rho \frac{\partial \bar{\psi}}{\partial \mathbf{a}_{\alpha}}. \tag{5.209}$$

Similar calculations, which we shall omit here, lead to

$$\mathbf{M}^{N\alpha} = \rho \frac{\partial \bar{\psi}}{\partial \mathbf{d}_{N}^{;\alpha}}, \tag{5.210}$$

$$\mathbf{m}^{N} = \rho \frac{\partial \bar{\psi}}{\partial \mathbf{d}_{N}}. \tag{5.211}$$

Equations (5.209)–(5.211) constitute an infinite set of equations that need to be complemented by an infinite number of initial and boundary conditions for the well-posedness of the problem. This drawback represents no improvement in the three-dimensional theory of continuum media and requires for a reduction to a finite set of equations and conditions. This reduction can be achieved either by means of a variational approach (see [Dikmen, 1982], Chapter 6) or by looking for instances in which one can approximate the problem by eliminating all quantities with $N \geq 1$.

5.7.1 *Linear constitutive equations*

A problem concerning the theory of elastic shells is said to be geometrically linear if it is assumed a linearized kinematics for the formulation of the problem. On the other hand, the problem is said to be physically linear if the stress-strain relations are linear. [Novozhilov, 1961] divided the problems of elasticity in four classes:

(1) Physically linear and geometrically linear.
(2) Physically nonlinear but geometrically linear.
(3) Physically linear but geometrically nonlinear.
(4) Physically nonlinear and geometrically nonlinear.

Here we address the class of the fully linear problems, that is, those that are both physically and geometrically linear, with isotropic behavior (where the relevant properties are the same in all directions). In this case, the strain-energy density function is given by the following quadratic form (see [Dikmen, 1982], p. 82):

$$\rho_o^* \psi^* = \frac{E}{2(1+\nu)} \Big(G^{il} G^{jk} + G^{ik} G^{jl} + \frac{2\nu}{1-2\nu} G^{ij} G^{lk} \Big) \gamma_{ik}^* \gamma_{ij}^*, \qquad (5.212)$$

where E and ν are constants. Using this quadratic form in (5.195) we get the Hooke's law

$$T^{ij} = \frac{E}{2(1+\nu)} \Big(G^{il} G^{jk} + G^{ik} G^{jl} + \frac{2\nu}{1-2\nu} G^{ij} G^{lk} \Big) \gamma_{lk}^* \qquad (5.213)$$

for the Cauchy stress tensor.

If we introduce the *complementary energy density* τ as a function of the components T^{ij} by means of the so-called Legendre transformation

$$\tau = -\psi^* + \frac{1}{\rho_o^*} T^{ij} \gamma_{ij}^* \qquad (5.214)$$

and differentiate it, we obtain

$$\frac{\partial \tau}{\partial T^{lk}} = \left(\frac{1}{\rho_o^*} T^{ij} - \frac{\partial \psi^*}{\partial \gamma_{ij}^*} \right) \frac{\partial \gamma_{ij}^*}{\partial T^{lk}} + \frac{1}{\rho_o^*} \frac{\partial T^{ij}}{\partial T^{lk}} \gamma_{ij}^*. \tag{5.215}$$

By virtue of (5.195), this reduces to

$$\rho_o^* \frac{\partial \tau}{\partial T^{lk}} = \gamma_{lk}^*. \tag{5.216}$$

Considering that we can equate $\tau(T^{ij})$ and $\psi(T^{ij})$, we may write (5.216) as (see Langhaar [Langhaar, 1962] and Reddy [Reddy, 2002])

$$\rho_o^* \frac{\partial \psi^*}{\partial T^{lk}} = \gamma_{lk}^*. \tag{5.217}$$

The isotropic behavior results in $\psi^*(T^{ij})$ being given by

$$\rho_o^* \psi^*(T^{ij}) = \rho_o^* \tau = \left(\frac{1+\nu}{2E} G_{il} G_{jk} - \frac{\nu}{2E} G_{ij} G_{lk} \right) T^{ij} T^{lk}. \tag{5.218}$$

Using the strain energy density per unit of area (5.208), a complementary energy density $\bar{\tau}$, or $\bar{\psi}^*(T^{ij})$, can be introduced by the formula

$$\bar{\tau} = \bar{\psi}^*(T^{ij}) = \frac{1}{\rho_o \sqrt{A}} \int_\alpha^\beta \rho_o^* \psi^*(T^{ij}) G^{\frac{1}{2}} d\xi. \tag{5.219}$$

With the purpose of expressing one set of constitutive equations in terms of physical components, we recall the equations of motion (5.180) in terms of tensor components relating the variables $N^{\alpha\beta}, M^{\alpha i}$, and V^i and the kinematic measures $e_{\alpha\beta}, \rho_{i\alpha}, \gamma_i$ and consider the constitutive equations for an isotropic material — presented in [Naghdi, 1972], p. 597, and reproduced below — which is obtained assuming the restriction that $\bar{\psi}$ is independent of $B_{\alpha\beta}$ and that

$$\bar{\psi}(e_{\alpha\beta}, \rho_{\alpha\beta}, \rho_{3\alpha}, \gamma_\alpha, \gamma_3; A_{\alpha\beta}) = \bar{\psi}(e_{\alpha\beta}, -\rho_{\alpha\beta}, \rho_{3\alpha}, -\gamma_\alpha, \gamma_3; A_{\alpha\beta}).$$

These restrictions yield that $\bar{\psi}$ is given by

$$\bar{\psi} = \bar{\psi}_e + \bar{\psi}_b, \tag{5.220}$$

with

$$2\varrho_o\bar{\psi}_e = [\alpha_1 A^{\alpha\beta}A^{\gamma\delta} + \alpha_2(A^{\alpha\gamma}A^{\beta\delta} + A^{\alpha\delta}A^{\beta\gamma})]e_{\alpha\beta}e_{\gamma\delta} + \alpha_4(\gamma_3)^2 + \alpha_8 A^{\alpha\beta}_{3\alpha}\rho_{3\alpha}\rho_{3\beta} + 2\alpha_9 A^{\alpha\beta}e_{\alpha\beta}\gamma_3,$$

$$2\varrho_o\bar{\psi}_b = [\alpha_5 A^{\alpha\beta}A^{\gamma\delta} + \alpha_6 A^{\alpha\gamma}A^{\beta\delta} + \alpha_7 A^{\alpha\delta}A^{\beta\gamma}]\rho_{\alpha\beta}\rho_{\gamma\delta} + \alpha_3 A^{\alpha\beta}\gamma_\alpha\gamma_\beta.$$

Now, applying the differentiations (see [Naghdi, 1972], p. 555)

$$N'^{\alpha\beta} = \rho_o\frac{\partial\bar{\psi}}{\partial e_{\alpha\beta}} = N'^{\beta\alpha}, \tag{5.221}$$

$$V^i = \rho_o\frac{\partial\bar{\psi}}{\partial\gamma_i}, \tag{5.222}$$

$$M^{\alpha i} = \rho_o\frac{\partial\bar{\psi}}{\partial\rho_{i\alpha}}, \tag{5.223}$$

the resulting constitutive equations become

$$N'^{\alpha\beta} = (\alpha_1 A^{\alpha\beta}A^{\gamma\delta} + \alpha_2(A^{\alpha\gamma}A^{\beta\delta} + A^{\alpha\delta}A^{\beta\gamma}))e_{\gamma\delta} + \alpha_9 A^{\alpha\beta}\gamma_3, \tag{5.224}$$

$$M^{\beta\alpha} = (\alpha_5 A^{\alpha\beta}A^{\gamma\delta} + \alpha_6 A^{\alpha\gamma}A^{\beta\delta} + \alpha_7 A^{\alpha\delta}A^{\beta\gamma})\rho_{\gamma\delta}, \tag{5.225}$$

$$V^\alpha = \alpha_3 A^{\alpha\gamma}\gamma_\gamma, \quad V^3 = \alpha_4\gamma_3 + \alpha_9 A^{\alpha\beta}e_{\alpha\beta}, \tag{5.226}$$

$$M^{\alpha 3} = \alpha_8 A^{\alpha\gamma}\rho_{3\gamma}. \tag{5.227}$$

We can express them in terms of physical components as follows:

$$\left.\begin{aligned}
[N'^{\alpha\beta}]_{F\otimes E} &= \alpha_1([A^{\alpha\beta}]_{F\otimes E}[A^{\gamma\delta}]_{F\otimes E})[e_{\gamma\delta}]_{F^*\otimes E^*}\\
&\quad +\alpha_2([A^{\alpha\gamma}]_{F\otimes F}[A^{\beta\delta}]_{E\otimes E})[e_{\gamma\delta}]_{F^*\otimes E^*}\\
&\quad +\alpha_2([A^{\alpha\delta}]_{F\otimes E}[A^{\beta\gamma}]_{E\otimes F})[e_{\gamma\delta}]_{F^*\otimes E^*}\\
&\quad + \alpha_9[A^{\alpha\beta}]_{F\otimes E}\,\gamma_3,\\
[M^{\beta\alpha}]_{E\otimes F} &= \alpha_5([A^{\alpha\beta}]_{F\otimes E}[A^{\gamma\delta}]_{F\otimes E})[\varrho_{\gamma\delta}]_{F^*\otimes E^*}\\
&\quad +\alpha_6([A^{\alpha\gamma}]_{F\otimes F}[A^{\beta\delta}]_{E\otimes E})[\varrho_{\gamma\delta}]_{F^*\otimes E^*}\\
&\quad +\alpha_7([A^{\alpha\delta}]_{F\otimes E}[A^{\beta\gamma}]_{E\otimes F})[\rho_{\gamma\delta}]_{F^*\otimes E^*},\\
[V^\alpha]_F &= \alpha_3[A^{\alpha\gamma}]_{F\otimes F}[\gamma_\gamma]_{F^*},\\
V^3 &= \alpha_4\gamma_3 + \alpha_9[A^{\alpha\beta}]_{F\otimes E}[e_{\alpha\beta}]_{F^*\otimes E^*},\\
[M^{\alpha 3}]_F &= \alpha_8[A^{\alpha\gamma}]_{F\otimes F}[\rho_{3\gamma}]_{F^*},
\end{aligned}\right\} \tag{5.228}$$

where, from Table 4.2, we can establish

$$[A^{\gamma\delta}]_{F\otimes E} = \frac{\sqrt{G_{\delta\delta}}}{\sqrt{G^{\gamma\gamma}}} A^{\gamma\delta}, \quad [\,A^{\gamma\gamma}]_{E\otimes F} = \frac{\sqrt{G_{\gamma\gamma}}}{\sqrt{G^{\delta\delta}}} A^{\gamma\delta}, \tag{5.229}$$

$$[e_{\gamma\delta}]_{F^*\otimes E^*} = \frac{\sqrt{G^{\gamma\gamma}}}{\sqrt{G_{\delta\delta}}} e_{\gamma\delta}, \quad [e_{\gamma\delta}]_{E^*\otimes F^*} = \frac{\sqrt{G^{\delta\delta}}}{\sqrt{G_{\gamma\gamma}}} e_{\gamma\delta}, \tag{5.230}$$

$$[\rho_{\gamma\delta}]_{F^*\otimes E^*} = \frac{\sqrt{G^{\gamma\gamma}}}{\sqrt{G_{\delta\delta}}} \rho_{\gamma\delta}, \quad [\rho_{\gamma\delta}]_{E^*\otimes F^*} = \frac{\sqrt{G^{\delta\delta}}}{\sqrt{G_{\gamma\gamma}}} \rho_{\gamma\delta}. \tag{5.231}$$

At this point, we have two alternatives: either we solve the system (5.224) through (5.227) for the unknown tensor components and substitute them into (5.229) through (5.231) to determine the physical components or else we solve the system (5.228) for the physical components and substitute them into (5.229) through (5.231) to determine the tensor components.

Bibliography

Altman, W. and Oliveira, A.M. *Physical components of tensors. Tensor, shells in terms of anholonomic components.* Appl. Mech. Rev. **46**(11), 92–104 (1993); **31**, 141–148 (1977).

Altman, W. and Oliveira, A.M. *A historical outline of physical components of tensors.* Tensor. **36**, 195–199 (1982).

Altman, W. and Oliveira, A.M. *Physical and anholonomic components of tensors of elasticity theory.* Int. J. NonLinear Mechanics. **30**(3), 341–358 (1995).

Altman, W. and Oliveira, A.M. *Physical and anholonomic components of tensors via invariance of the tensor representation.* Mechanics of Advanced Materials and Structures. **18**, 454–466 (2011).

Amaral, L.H. *Linear Algebra and Geometry.* Biblioteca Nacional, São José dos Campos SP, Brazil (2002).

Bowen, R.M. and Wang, C.C. *Linear and Multilinear Algebra Introduction to Vectors and Tensors.* Vols. 1 and 2. Basic Books (1976).

Cartan, E. *Geometry of Riemannian spaces.* In Lie Group: History, Frontiers and Applications - Series A, volume XIII. Math Sci Press, Brookline, MA (1983).

Chung, T.J. *Continuum Mechanics.* Prentice–Hall, Englewood Cliffs, NJ (1988).

Cissotti, U. *Lezione di calcolo tensoriale.* Libreria Ed. Politecnica (1928).

Cosserat, E. and Cosserat, F. *Sur la théorie des corps minces.* Compt. Rend. **146**, 169–172 (1908).

Cosserat, E. and Cosserat, F. *Théorie des Corps Déformable,* 2nd ed. Appendix, pp. 953–1173 of Chwolson's Traité de Physique, Paris (1909).

Cotton, E. *Genéralisation de la theorie du trièdre mobile.* Bull. Soc. Math. **33**, 1–23 (1905).

Dikmen, M. *Theory of Thin Elastic Shells.* Pitman Publishing Inc., Mansfield, MA (1982).

Ericksen, J.L. *Tensor Fields.* Handbuch der Physik. Band III/1. Springer, Berlin (1960).

Ericksen, J.L. and Truesdell, C. *Exact theory and stress and strain in rods and shells.* Arch. Rational Mech. Anal. **1**, 295–323 (1958).

Green, A.E. and Zerna, W. *Theoretical Elasticity.* Clarendon Press, Oxford (1954).

Gurtin, M. E. *An Introduction to Continuum Mechanics.* Academic Press, New York (1982).

HENCKY, H. *Die bewegungsgleichungen beim nichtstationaren fliessen plastischer massen.* Zeitschrift für Angewandte Mathematik und Mechanik. **5**, 144–146 (1925).

HOFFMAN, K. AND KUNZE, R. *Linear Algebra.* Prentice–Hall, Engelwood Cliffs, NJ (1971).

HORN, R.A. AND JOHNSON, C.R. *Matrix Analysis.* Cambridge University Press, New York (1985).

KIRCHHOFF, G. *Über das Gleichgewicht und die Bewegung einer elastischen Scheibe.* Crelles J. **40**, 51–88; Ges. Abh. 237–279 (1850).

LANGHAAR, H.L. *Energy Methods in Applied Mechanics.* John Wiley, New York (1962).

LIPSCHUTZ, S. *Schaum's Outline of Theory and Problems of Linear Algebra.* Schaum Publishing (1968).

LOVE, A.E.H. *The small free vibrations and deformation of a thin elastic shell.* Phil. Trans. Roy. London, Ser. A **179**, 491–546 (1888).

MCCONNELL, A.J. *Applications of Absolute Differential Calculus.* Blackie, London (1931).

NAGHDI, P.M. *On the theory of thin elastic shells.* Quart. Appl. Math. **14** , 369–380 (1957).

NAGHDI, P.M. *The Theory of Shells and Plates.* In Flügge, S. & Truesdell, C. (eds.): Handbuch der Physik, Band VI a/2, pp. 425–640. Springer, Berlin (1972).

NOVOZHILOV, V. *The Theory of Thin Shells.* (Translated from the 1951 Russian ed.) P. Noordhoff (1970).

NOVOZHILOV, V. *The Theory of Elasticity.* Israel Programfor Scientific Translation. Jerusalem (1961).

NOVOZHILOV, V. AND FINKELSHTEIN, R. *On the incorrectness of Kirchhoff's hypotheses in the theory of shells* [in Russian]. Prikl. Mat. Mech. **7**, 331–340 (1943).

OLIVEIRA, A.M. AND ALTMAN, W. *Anholonomic components of vectors and tensors.* Tensor, N.S. **35**, 283–286 (1981).

REDDY, J.N., *Energy Principles and Variational Methods in Applied Mechanics,* 2nd ed., John Wiley & Sons, New York (2002).

REDDY, J.N., *Mechanics of Laminated Composite Plates and Shells,* 2nd ed., CRC Press, Boca Raton, FL (2004).

REDDY, J.N., *Theory and Analysis of Elastic Plates and Shells,* 2nd ed., CRC Press, Boca Raton, FL (2007).

REDDY, J.N., *An Introduction to Continuum Mechanics,* 2nd ed., Cambridge University Press, New York (2013).

REDDY, J.N. AND RASMUSSEN, L. *Advanced Engineering Analysis.* John Wiley, New York (1982); reprinted by Krieger (1992).

REISSNER, E. *A new derivation of the equations for the deformation of elastic shells.* Amer. J. Math. **63**, 177–184 (1941).

RICCI, G. AND LEVI-CIVITA, T.*Méthodes du calcul differéntiel absolu et leurs applications.* Math. Annalen. **54**, 125–201 (1901).

SCHOUTEN, J.A. *Tensor Analysis for Physicists.* Clarendon Press, Oxford (1951).

SEDOV, L.I. *Introduction to the Mechanics of Continuous Medium.* Addison-Wesley, London (1965).

SOKOLNIKOFF, I.S. *Tensor Analysis*, 2nd ed., John Wiley & Sons, New York (1964).

SYNGE, J.L. AND CHIEN, W.Z. *The intrinsic theory of elastic shells and plates.* Th. von Kármán Anniv. Volume, 103–120 (1941).

SYNGE, J.L. AND SCHILD, A. *Tensor Calculus.* Toronto University Press, Toronto, 1949. Republished by Dover, New York (1978).

THIRRING, H. *Zur tensoranalytischen darstellung der elastizitltstheorie.* Phys. Z. **26**, 518–522 (1925).

TRUESDELL, C. *The physical components of vectors and tensors.* Zangew Math. Mech. **33**, 345–356 (1953).

TRUESDELL, C. *Remarks on the paper "The physical components of vectors and tensors."* Zangew Math. Mech. **34**, 69–70 (1954).

TRUESDELL, C. AND TOUPIN, R.A. *The classical field theories.* In Fl ügge, S. (ed.): Handbuch der Physik III/1, pp. 226–793. Springer, Berlin, 1960.

ZERNA, W. *A new formulation of the theory of elastic shells.* IASS Bull. **37**, 61–76 (1968).

Index

Printed in the United States
by Baker & Taylor Publisher Services